Everyday Mathematics®

The University of Chicago School Mathematics Project

Study Links

Grade **5**

McGraw Hill **Wright Group**

The McGraw-Hill Companies

The University of Chicago School Mathematics Project (UCSMP)

Max Bell, Director, UCSMP Elementary Materials Component; Director, *Everyday Mathematics* First Edition; James McBride, Director, *Everyday Mathematics* Second Edition; Andy Isaacs, Director, *Everyday Mathematics* Third Edition; Amy Dillard, Associate Director, *Everyday Mathematics* Third Edition

Authors

Max Bell, John Bretzlauf, Amy Dillard, Robert Hartfield, Andy Isaacs, James McBride, Kathleen Pitvorec, Peter Saecker, Noreen Winningham*, Robert Balfanz†, William Carroll†

*Third Edition only   †First Edition only

Technical Art
Diana Barrie

Teachers in Residence
Fran Goldenberg,
Sandra Vitantonio

Editorial Assistant
Rosina Busse

Contributors

Tammy Belgrade, Diana Carry, Debra Dawson, Kevin Dorken, James Flanders, Laurel Hallman, Ann Hemwall, Elizabeth Homewood, Linda Klaric, Lee Kornhauser, Judy Korshak-Samuels, Deborah Arron Leslie, Joseph C. Liptak, Sharon McHugh, Janet M. Meyers, Susan Mieli, Donna Nowatzki, Mary O'Boyle, Julie Olson, William D. Pattison, Denise Porter, Loretta Rice, Diana Rivas, Michelle Schiminsky, Sheila Sconiers, Kevin J. Smith, Teresa Sparlin, Laura Sunseri, Kim Van Haitsma, John Wilson, Mary Wilson, Carl Zmola, Teresa Zmola

Photo Credits

©Burke/Triolo Productions/Getty Images, p. v; ©C Squared Studios/Getty Creative, p. iii; ©Chicago Educational Publishers, p. 141; ©W. Perry Conway/CORBIS, cover, *right*; ©Food Collection/Index Stock, p. 71; ©Getty Images, cover, *bottom left*; ©Ken O'Donoghue, pp. 23, 89, 117, 219, 241; ©PIER/Getty Images, cover, *center*.

Permissions

Excerpt from "Arithmetic" in THE COMPLETE POEMS OF CARL SANDBURG, copyright © 1970, 1969 by Lilian Steichen Sandburg, Trustee, reprinted by permission of Harcourt, Inc. This material may not be produced in any form or by any means without the prior written permission of the publisher; Excerpt from "Finding Time" by JoAnne Growney, Mathematics Magazine, vol. 68, no. 4 (October 1995); *Second Poem: "123"* by Ken Stange, first published in *Cold Pigging Poetics (Hypothesis 5)*, ISBN 0-920424-25-2, 1981, York Publishing, Toronto.

This material is based upon work supported by the National Science Foundation under Grant No. ESI-9252984. Any opinions, findings, conclusions, or recommendations expressed in this material are those of the authors and do not necessarily reflect the views of the National Science Foundation.

www.WrightGroup.com

Wright Group

Send all inquiries to:
Wright Group/McGraw-Hill
P.O. Box 812960
Chicago, IL 60681

ISBN-13 978-0-07-609742-5
ISBN-10 0-07-609742-0

4 5 6 7 8 9 DBH 12 11 10 09 08 07

The McGraw·Hill Companies

Contents

Contents   **v**

STUDY LINK
1·1

Number Poetry

Many poems have been written about mathematics. They are poems that share some of the ways that poets think about numbers and patterns.

1. Read the examples below.

2. The ideas in the examples are some of the ideas you have studied in *Everyday Mathematics.* Subtraction is one of these ideas. Name as many other ideas from the examples as you can on the back of this page.

Examples:

Arithmetic is where numbers fly like pigeons in and out of your head.
Arithmetic tells you how many you lose or win if you know how many you had before you lost or won.

                 from "Arithmetic" by Carl Sandburg

A square is neither a line
nor circle; it is timeless.
Points don't chase around
a square. Firm, steady,
it sits there and knows
its place. A circle
won't be squared.

             from "Finding Time" by JoAnne Growney

Second Poem: "123"

.
1
12
123
1-32
1-21
1-10

How many seconds in an hour?   2
How many in a day?   21
What size are the planets in the sky?   21-31
How far to the Milky Way?   2131
                 21-31-231
How fast does lightning travel?   121
How slow do feathers fall?   1
How many miles to Istanbul?
Mathematics knows it all!   .

      from "Marvelous Math" by Rebecca Kai Dotlich

from "Asparagus X Plus Y"
by Ken Stange

3. Use a number pattern to make your own poem on the back of this page.

1

STUDY LINK 1·1

Unit 1: Family Letter

Introduction to *Fifth Grade Everyday Mathematics*

Welcome to *Fifth Grade Everyday Mathematics*. This curriculum was developed by the University of Chicago School Mathematics Project to offer students a broad background in mathematics.

The features of the program described below are to help familiarize you with the structure and expectations of *Everyday Mathematics*.

A problem-solving approach based on everyday situations Students learn basic math skills in a context that is meaningful by making connections between their own knowledge and experience and mathematics concepts.

Frequent practice of basic skills Students practice basic skills in a variety of engaging ways. In addition to completing daily review exercises covering a variety of topics and working with multiplication and division fact families in different formats, students play games that are specifically designed to develop basic skills.

An instructional approach that revisits concepts regularly Lessons are designed to take advantage of previously learned concepts and skills and to build on them throughout the year.

A curriculum that explores mathematical content beyond basic arithmetic Mathematics standards around the world indicate that basic arithmetic skills are only the beginning of the mathematical knowledge students will need as they develop critical-thinking skills. In addition to basic arithmetic, *Everyday Mathematics* develops concepts and skills in the following topics—number and numeration; operations and computation; data and chance; geometry; measurement and reference frames; and patterns, functions, and algebra.

Everyday Mathematics provides you with ample opportunities to monitor your child's progress and to participate in your child's mathematical experiences. Throughout the year, you will receive Family Letters to keep you informed of the mathematical content your child is studying in each unit. Each letter includes a vocabulary list, suggested Do-Anytime Activities for you and your child, and an answer guide to selected Study Link (homework) activities.

Please keep this Family Letter for reference as your child works through Unit 1.

Fifth Grade Everyday Mathematics emphasizes the following content:

Number and Numeration Understand the meanings, uses, and representations of numbers; equivalent names for numbers, and common numerical relations.

Operations and Computation Make reasonable estimates and accurate computations; understand the meanings of operations.

Data and Chance Select and create appropriate graphical representations of collected or given data; analyze and interpret data; understand and apply basic concepts of probability.

Geometry Investigate characteristics and properties of 2- and 3-dimensional shapes; apply transformations and symmetry in geometric situations.

Measurement and Reference Frames Understand the systems and processes of measurement; use appropriate techniques, tools, units, and formulas in making measurements; use and understand reference frames.

Patterns, Functions, and Algebra Understand patterns and functions; use algebraic notation to represent and analyze situations and structures.

Unit 1: Number Theory

In Unit 1, students study properties of whole numbers by building on their prior work with multiplication and division of whole numbers.

Students will collect examples of arrays to form a class Arrays Museum. To practice using arrays with your child at home, use any small objects, such as beans, macaroni, or pennies.

Building Skills through Games

In Unit 1, your child will practice operations and computation skills by playing the following games. Detailed instructions for each game are in the *Student Reference Book*.

Factor Bingo This game involves 2 to 4 players and requires a deck of number cards with 4 each of the numbers 2–9, a drawn or folded 5-by-5 grid and 12 pennies or counters for each player. The goal of the game is to practice the skill of recognizing factors.

Factor Captor See *Student Reference Book*, page 306. This is a game for 2 players. Materials needed include a *Factor Captor* Grid, 48 counters the size of a penny, scratch paper, and a calculator. The

goal of the game is to strengthen the skill of finding the factors of a number.

Multiplication Top-It See *Student Reference Book*, page 334. This game requires a deck of cards with 4 each of the numbers 1–10 and can be played by 2–4 players. *Multiplication Top-It* is used to practice the basic multiplication facts.

Name That Number See *Student Reference Book*, page 325. This game involves 2 or 3 players and requires a complete deck of number cards. *Name That Number* provides practice with computation and strengthens skills related to number properties.

Vocabulary

Important terms in Unit 1:

composite number A counting number greater than 1 that has more than two *factors*. For example, 4 is a composite number because it has three factors: 1, 2, and 4.

divisible by If the larger of two counting numbers can be divided by the smaller with no remainder, then the larger is divisible by the smaller. For example, 28 is divisible by 7 because $28 / 7 = 4$ with no remainder.

exponent The small, raised number in exponential notation that tells how many times the base is used as a *factor*.

Example:

$5^2 \leftarrow$ exponent $\quad 5^2 = 5 * 5 = 25.$

$10^3 \leftarrow$ exponent $\quad 10^3 = 10 * 10 * 10 = 1,000.$

$2^4 \leftarrow$ exponent $\quad 2^4 = 2 * 2 * 2 * 2 = 16.$

factor One of two or more numbers that are multiplied to give a *product*.

$3 * 5 = 15$ $\qquad$ $15 * 1 = 15$

Factors Product $\qquad$ Factors Product

factor rainbow A way to show factor pairs in a list of all the factors of a number. A factor rainbow can be used to check whether a list of factors is correct.

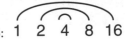

Factor rainbow for 16: 1   2   4   8   16

number model A number sentence or expression that models a number story or situation. For example, a number model for the array below is $4 * 3 = 12$.

prime number A whole number that has exactly two factors: itself and 1. For example, 5 is a prime number because its only factors are 5 and 1.

product The result of multiplying two or more numbers, called *factors*.

rectangular array A rectangular arrangement of objects in rows and columns such that each row has the same number of objects and each column has the same number of objects.

square number A number that is the product of a counting number multiplied by itself. For example, 25 is a square number, because $25 = 5 * 5$.

As You Help Your Child with Homework

As your child brings assignments home, you might want to go over the instructions together, clarifying them as necessary. The answers listed below will guide you through this unit's Study Links.

Study Link 1·2

1. $1 * 5 = 5$
$5 * 1 = 5$

2. $1 * 14 = 14$; $14 * 1 = 14$
$2 * 7 = 14$; $7 * 2 = 14$

3. $1 * 18 = 18$; $18 * 1 = 18$; $2 * 9 = 18$;
$9 * 2 = 18$; $3 * 6 = 18$; $6 * 3 = 18$

4. 795   **5.** 271   **6.** 98   **7.** 984   **8.** 5

Study Link 1·3

1.   24; 24

3.   24; 3, 8; 24

6. $1 * 5 = 5$; 1, 5

7. 4   **8.** 3,919   **9.** 2,763   **10.** 159

Study Link 1·4

1. The next number to try is 5, but 5 is already listed as a factor. Also, any factor greater than 5 would already be named because it would be paired with a factor less than 5.

2. 1, 5, 25   **3.** 1, 2, 4, 7, 14, 28

4. 1, 2, 3, 6, 7, 14, 21, 42

5. 1, 2, 4, 5, 10, 20, 25, 50, 100

6. 9,551   **7.** 48   **8.** 41,544   **9.** 441   **10.** 7

Study Link 1·5

1. Divisible by 2: 998,876; 5,890; 36,540; 1,098
Divisible by 3: 36,540; 33,015; 1,098
Divisible by 9: 36,540; 1,098
Divisible by 5: 5,890; 36,540; 33,015

2. Divisible by 4: 998,876; 36,540

3. 1,750   **4.** 8,753   **5.** 250   **6.** 13

Study Link 1·6

1. 11; 1, ⑪;   p
2. 18; 1, ②③ 6, 9, 18;   c
3. 24; 1, ②③ 4, 6, 8, 12, 24;   c
4. 28; 1, ② 4, ⑦ 14, 28;   c
5. 36; 1, ②③ 4, 6, 9, 12, 18, 36;   c
6. 49; 1, ⑦ 49;   c
7. 50; 1, ②⑤ 10, 25, 50;   c
8. 70; 1, ②⑤⑦ 10, 14, 35, 70;   c
9. 100; 1, ② 4, ⑤ 10, 20, 25, 50, 100;   c

10. 9,822   **11.** 234   **12.** 21,448   **13.** 9 R3

Study Link 1·7

1. 16   **2.** 49   **3.** 6   **4.** 64   **5.** 25

6. 81   **7.** $4 * 9 = 36$ **8.** $5 * 5 = 25$

9. a. $5 * 5 = 25$

   b. $5 * 5 = 25$ shows a square number because there are the same number of rows and columns. A square can be drawn around this array.

Study Link 1·8

1. 36: 1, 2, 3, 4, 6, 9, 12, 18, 36; $6^2 = 36$ The square root of 36 is 6.

3. $11^2 = 121$; the square root of 121 is 11.

5. 6,219   **6.** 3,060   **8.** 8 R2   **9.** 42

Study Link 1·9

1. b. $7^2 = 7 * 7 = 49$

   c. $20^3 = 20 * 20 * 20 = 8,000$

2. a. 11^2 **b.** 9^3 **c.** 50^4

3. a. $2 * 3^3 * 5^2 = 2 * 3 * 3 * 3 * 5 * 5 = 1,350$

   b. $2^4 * 4^2 = 2 * 2 * 2 * 2 * 4 * 4 = 256$

4. a. $40 = 2 * 2 * 2 * 5 = 2^3 * 5$

   b. $90 = 2 * 3 * 3 * 5 = 2 * 3^2 * 5$

5. 5,041   **6.** 720   **7.** 50 R4   **8.** 99,140

9. 12   **10.** 47,668

STUDY LINK 1·2 | More Array Play

A **rectangular array** is an arrangement of objects in rows and columns. Each row has the same number of objects, and each column has the same number of objects. We can write a multiplication number model to describe a rectangular array.

● ● ●
● ● ●
● ● ●
● ● ●

4 * 3 = 12

For each number below, use pennies or counters to make as many different arrays as possible. Draw each array on the grid with dots. Write the number model next to each array.

1. 5

2. 14

3. 18

Practice

4. 487 + 308 = _____

5. 679 − 408 = _____

6. 14 * 7 = _____

7. 164 * 6 = _____

8. 45 ÷ 9 = _____

Factors

To find the factors of a number, ask yourself: *Is 1 a factor of the number?*
Is 2 a factor? Is 3 a factor? Continue with larger numbers. For example, to
find all the factors of 15, ask yourself these questions.

	Yes/No	Number Sentence	Factor Pair
Is 1 a factor of 15?	Yes	1 * 15 = 15	1, 15
Is 2 a factor of 15?	No		
Is 3 a factor of 15?	Yes	3 * 5 = 15	3, 5
Is 4 a factor of 15?	No		

1. You don't need to go any further. Can you tell why?

_____

_____

_____

_____

So the factors of 15 are 1, 3, 5, and 15.

List as many factors as you can for each of the numbers below.

2. 25 _____

3. 28 _____

4. 42 _____

5. 100 _____

Practice

6. 8,417 + 1,134 = _____

7. 73 − 25 = _____

8. 6,924 * 6 = _____

9. 634 − 193 = _____

10. 56 / 8 = _____

STUDY LINK 1·5 | Divisibility Rules

All even numbers are divisible by 2.

A number is divisible by 3 if the sum of its digits is divisible by 3.

A number is divisible by 6 if it is divisible by both 2 and 3.

A number is divisible by 9 if the sum of its digits is divisible by 9.

A number is divisible by 5 if it ends in 0 or 5.

A number is divisible by 10 if it ends in 0.

1. Use divisibility rules to test whether each number is divisible by 2, 3, 5, 6, 9, or 10.

Number	Divisible...					
	by 2?	by 3?	by 6?	by 9?	by 5?	by 10?
998,876						
5,890						
36,540						
33,015						
1,098						

A number is divisible by 4 if the tens and ones digits form a number that is divisible by 4.

Example: 47,8**36** is divisible by 4 because 36 is divisible by 4.

It isn't always easy to tell whether the last two digits form a number that is divisible by 4. A quick way to check is to divide the number by 2 and then divide the result by 2. It's the same as dividing by 4, but is easier to do mentally.

Example: 5,384 is divisible by 4 because 84 / 2 = 42 and 42 / 2 = 21.

2. Place a star next to any number in the table that is divisible by 4.

Practice	

3. 250 * 7 = _____

4. 1,931 + 4,763 + 2,059 = _____

5. (20 + 30) * 5 = _____

6. 78 ÷ 6 = _____

Prime and Composite Numbers

A **prime number** is a whole number that has exactly two factors—1 and the number itself. A **composite number** is a whole number that has more than two factors.

For each number:

◆ List all of its factors.

◆ Write whether the number is prime or composite.

◆ Circle all of the factors that are prime numbers.

	Number	Factors	Prime or Composite?
1	11		
2	18		
3	24		
4	28		
5	36		
6	49		
7	50		
8	70		
9	100		

Practice

10. $4,065 + 2,803 + 2,954 =$ _____

11. $392 - 158 =$ _____

12. $1,532 * 14 =$ _____

13. $39 / 4 \rightarrow$ _____

14. $48 * 15 =$ _____

Exploring Square Numbers

A **square number** is a number that can be written as the product of a number multiplied by itself. For example, the square number 9 can be written as 3 * 3.

$9 = 3 * 3 = 3^2$

Fill in the missing numbers.

1. $4 * 4 =$ _____

2. _____ $= 7 * 7$

3. _____ $* 6 = 36$

4. $8^2 =$ _____

5. $5^2 =$ _____

6. _____ $= 9^2$

Write a number model to describe each array.

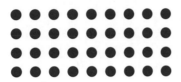

7. Number model: _____

8. Number model: _____

9. a. Which of the arrays above shows a square number? _____

b. Explain your answer.

_____

_____

_____

Practice

10. $97 * 43 =$ _____

11. $4,006 - 2,675 =$ _____

12. $1,416 + 8,348 =$ _____

13. $725 - 414 =$ _____

17

STUDY LINK
1·8

Factor Rainbows, Squares, and Square Roots

1. List all the factors of each square number. Make a **factor rainbow** to check your work. Then fill in the missing numbers.

Reminder: In a factor rainbow, the product of each connected factor pair should be equal to the number itself. For example, the factor rainbow for 16 looks like this:

1 2 4 8 16

$1 * 16 = 16$ $2 * 8 = 16$ $4 * 4 = 16$

Example:

4: *1, 2, 4* $\widehat{1\ 2\ 4}$

$\underline{2}^2 = 4$ The square root of 4 is *2*.

9:

$\underline{}^2 = 9$ The square root of 9 is ___.

25:

$\underline{}^2 = 25$ The square root of 25 is ___.

36:

$\underline{}^2 = 36$ The square root of 36 is ___.

---

2. Do all square numbers have an odd number of factors? _____

Unsquare each number. The result is its square root. Do not use the square root key $\boxed{\sqrt{}}$ on your calculator.

3. $\underline{}^2 = 121$

The square root of 121 is _____.

4. $\underline{}^2 = 2,500$

The square root of 2,500 is _____.

Practice

5.   4,318
      + 1,901

6.   36
        × 85

7.   2,852
      ×   5

8. $50 \div 6 \rightarrow$ _____

9. $333 - 291 =$ _____

STUDY LINK 1·9 | Exponents

An **exponent** is a raised number that shows how many times the number to its left is used as a factor.

Examples: 5^2 ← exponent 5^2 means 5 * 5, which is 25.
   10^3 ← exponent 10^3 means 10 * 10 * 10, which is 1,000.
   2^4 ← exponent 2^4 means 2 * 2 * 2 * 2, which is 16.

1. Write each of the following as a factor string. Then find the product.

   Example: 2^3 = $\underline{2*2*2}$ = $\underline{8}$ **a.** 10^4 = _____ = _____

   b. 7^2 = _____ = _____   **c.** 20^3 = _____ = _____

2. Write each factor string using an exponent.

   Example: 6 * 6 * 6 * 6 = $\underline{6^4}$ **a.** 11 * 11 = _____

   b. 9 * 9 * 9 = _____   **c.** 50 * 50 * 50 * 50 = _____

3. Write each of the following as a factor string that does *not* have any exponents. Then use your calculator to find the product.

   Example: 2^3 * 3 = $\underline{2*2*2*3}$ = $\underline{24}$

   a. $2 * 3^3 * 5^2$ = _____ = _____

   b. $2^4 * 4^2$ = _____ = _____

4. Write the prime factorization of each number. Then write it using exponents.

   Example: 18 = $\underline{2 * 3 * 3}$ = $\underline{2 * 3^2}$

   a. 40 = _____ = _____

   b. 90 = _____ = _____

| **Practice** |

5. 6,383 − 1,342 = _____   6. 48 * 15 = _____

7. 7)‾354‾ → _____   8. 50,314 + 48,826 = _____

9. 84 ÷ 7 = _____   10. 701 * 68 = _____

21

STUDY LINK
1·10

Unit 2: Family Letter

Estimation and Calculation

Computation is an important part of problem solving. Many of us were taught that there is just one way to do each kind of computation. For example, we may have learned to subtract by borrowing, without realizing that there are many other methods of subtracting numbers.

In Unit 2, students will investigate several methods for adding, subtracting, and multiplying whole numbers and decimals. Students will also take on an Estimation Challenge in Unit 2. For this extended problem, they will measure classmates' strides, and find a median length for all of them. Then they will use the median length to estimate how far it would take to walk to various destinations.

Throughout the year, students will practice using estimation, calculators, as well as mental and paper-and-pencil methods of computation. Students will identify which method is most appropriate for solving a particular problem. From these exposures to a variety of methods, they will learn that there are often several ways to accomplish the same task and achieve the same result. Students are encouraged to solve problems by whatever method they find most comfortable.

Computation is usually not the first step in the problem-solving process. One must first decide what numerical data are needed to solve the problem and which operations need to be performed. In this unit, your child will continue to develop his or her problem-solving skills with a special focus on writing and solving equations for problems.

Please keep this Family Letter for reference as your child works through Unit 2.

23

Vocabulary

Important terms in Unit 2:

Estimation Challenge A problem for which it is difficult, or even impossible, to find an exact answer. Your child will make his or her best estimate and then defend it.

magnitude estimate A rough estimate. A magnitude estimate tells whether an answer should be in the tens, hundreds, thousands, and so on.

Example: Give a magnitude estimate for 56 * 32

Step 1: Round 56 to 60.

Step 2: Round 32 to 30.

60 * 30 = 1,800, so a magnitude estimate for 56 * 32 is in the thousands.

| 10s | 100s | (1,000s) | 10,000s |

maximum The largest amount; the greatest number in a set of data.

mean The sum of a set of numbers divided by the number of numbers in the set. The mean is often referred to simply as the average.

median The middle value in a set of data when the data are listed in order from smallest to largest or vice versa. If there is an even number of data points, the median is the *mean* of the two middle values.

minimum The smallest amount; the smallest number in a set of data.

partial-sums addition A method, or algorithm, for adding in which sums are computed for each place (ones, tens, hundreds, and so on) separately and are then added to get a final answer.

```
                    268
                  + 483
1. Add 100s         600
2. Add 10s          140
3. Add 1s         +  11
4. Add partial sums. 751
```

Partial-sums algorithm

place value A number system that values a digit according to its position in a number. In our number system, each place has a value ten times that of the place to its right and one-tenth the value of the place to its left. For example, in the number 456, the 4 is in the hundreds place and has a value of 400.

range The difference between the *maximum* and *minimum* in a set of data.

reaction time The amount of time it takes a person to react to something.

trade-first subtraction A method, or algorithm, for subtracting in which all trades are done before any subtractions are carried out.

Example: 352 − 164

100s	10s	1s		100s	10s	1s
	4	12			14	
3	5̶	2̶		2	4̶	12
−1	6	4		3̶	5̶	2̶
				−1	6	4
				1	8	8

Trade 1 ten for 10 ones. Trade 1 hundred for 10 tens and subtract in each column.

24

Building Skills through Games

In Unit 2, your child will practice computation skills by playing these games. Detailed instructions are in the *Student Reference Book*.

Addition Top-It See *Student Reference Book,* page 333. This game for 2 to 4 players requires a calculator and 4 each of the number cards 1–10, and provides practice with place–value concepts and methods of addition.

High-Number Toss See *Student Reference Book,* pages 320 and 321. Two players need one six-sided die for this game. *High-Number Toss* helps students review reading, writing, and comparing decimals and large numbers.

Multiplication Bull's-Eye See *Student Reference Book,* page 323. Two players need 4 each of the number cards 0–9, a six-sided die, and a calculator to play this game. *Multiplication Bull's Eye* provides practice in estimating products.

Number Top-It See *Student Reference Book,* page 326. Two to five players need 4 each of the number cards 0–9 and a Place-Value Mat. Students practice making large numbers.

Subtraction Target Practice See *Student Reference Book,* page 331. One or more players need 4 each of the number cards 0–9 and a calculator. In this game, students review subtraction with multidigit whole numbers and decimals.

Do-Anytime Activities

To work with your child on the concepts taught in Units 1 and 2, try these activities:

1. When your child adds or subtracts multidigit numbers, talk about the strategy that works best. Try not to impose the strategy that works best for you! Here are some problems to try:

$467 + 343 =$ _____   _____ $= 761 + 79$

$894 - 444 =$ _____ $842 - 59 =$ _____

2. As you encounter numbers while shopping or on license plates, ask your child to read the numbers and identify digits in various places—thousands place, hundreds place, tens place, ones place, tenths place, and hundredths place.

As You Help Your Child with Homework

As your child brings assignments home, you might want to go over the instructions together, clarifying them as necessary. The answers listed below will guide you through this unit's Study Links.

Study Link 2·1

Answers vary for Problems 1–5.

6. 720 **7.** 90,361 **8.** 12 **9.** 18

Study Link 2·2

Sample answers:

1. 571 and 261 **2.** 30, 20, and 7

3. 19 and 23 **4.** 533 and 125

5. 85.2 and 20.5, or 88.2 and 17.5; Because the sum has a 7 in the tenths place, look for numbers with tenths that add to 7: 85.2 + 20.5 = 105.7; and 88.2 + 17.5 = 105.7.

6. 4,572 **7.** 4.4 **8.** 246 **9.** 1.918

10. 47 **11.** 208 **12.** 3 **13.** 8 R2

Study Link 2·3

1. 451 and 299 **2.** 100.9 and 75.3

3. Sample answer: 803 and 5,000

4. 17 and 15 **5.** 703 and 1,500

6. 25 and 9 **7.** 61 **8.** 137 **9.** 5.8

10. 18.85 **11.** 6 **12.** 84,018 **13.** $453.98

14. 98 **15.** 14

Study Link 2·4

1. a. 148 and 127 **b.** Total number of cards

   c. 148 + 127 = b **d.** b = 275

   e. 275 baseball cards

2. a. 20.00; 3.89; 1.49 **b.** The amount of change

   c. 20.00 − 3.89 − 1.49 = c,
      or 20 − (3.89 + 1.49) = c

   d. c = 14.62 **e.** $14.62

3. a. 0.6; 1.15; 1.35; and 0.925

   b. The length of the ribbons

   c. b = 0.6 + 1.15 + 1.35 + 0.925

   d. b = 4.025 **e.** 4.025 meters

Study Link 2·5

Answers vary for Problems 1–5.

6. 5,622 **7.** 29,616 **8.** 518 **9.** 13

Study Link 2·6

1. Unlikely: 30%   Very likely: 80%
   Very unlikely: 15%   Likely: 70%
   Extremely unlikely: 5%

2. 30%: Unlikely   5%: Extremely unlikely
   99%: Extremely unlikely   20%: Very unlikely
   80%: Very likely   35%: Unlikely
   65%: Likely   45%: 50-50 chance

Study Link 2·7

1. 1,000s; 70 ∗ 30 = 2,100

2. 1,000s; 10 ∗ 700 = 7,000

3. 10,000s; 100 ∗ 100 = 10,000

4. 10s; 20 ∗ 2 = 40

5. 10s; 3 ∗ 4 = 12

6. Sample answers: 45 ∗ 68 = 3,060;
   684 ∗ 5 = 3,420; and 864 ∗ 5 = 4,320

Study Link 2·8

1. 152; 100s; 8 ∗ 20 = 160

2. 930; 100s; 150 ∗ 6 = 900

3. 2,146; 1,000s; 40 ∗ 60 = 2,400

4. 21; 10s; 5 ∗ 4 = 20.

5. 26.04; 10s; 9 ∗ 3 = 27

Study Link 2·9

1. 6,862; 1,000s **2.** 88.8; 10s **3.** 33.372; 10s

4. 100,224; 100,000s **5.** 341.61; 100s

6. 9,989 **7.** 5 R2 **8.** 91 **9.** $19.00

Study Link 2·10

1. 390.756 **2.** 3,471.549 **3.** 9,340

4. 244 **5.** 44,604 **6.** 19 R2

STUDY LINK
2·1

Estimation

SRB
247

Class Medians for: Step Length _____ Steps in 1 Minute _____

A group of fifth-grade students in New Zealand are going camping. They will hike from Wellington to Ruapehu. Then they will follow a trail for another $\frac{1}{2}$ mile to their campsite. Use the map on this page (Scale: 1 inch = 400 miles) as well as your class median step length, and number of steps in 1 minute, to make the following estimates. (*Reminder:* 1 mile = 5,280 feet)

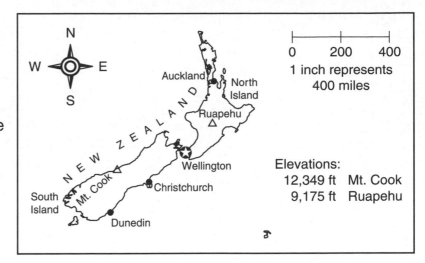

1. About how many miles is it from Wellington to Ruapehu? _____
   (unit)

2. About how many miles is it from Wellington to the campsite? _____
   (unit)

3. About how long would it take the students to arrive at their campsite, if they don't make any stops? _____
   (unit)

4. Each day, the students will hike for 12 hours and take 12 hours for stops to eat, rest, and sleep. If they leave at 7:00 A.M. on a Monday morning, at about what time, and on what day would you expect them to arrive at their campsite?

   Time: About _____ Day: _____

Try This

5. Suppose the students take a bus from Wellington to Mt. Cook and then hike to a campsite at the top of the mountain. Would they have to hike more or less than the distance they hiked to their campsite at Ruapehu?

Practice

6. 48 * 15 = _____

7. 24,029 + 26,840 + 39,492 = _____

8. 36 / 3 = _____

9. 35 − 17 = _____

STUDY LINK 2·2 | Number Hunt

SRB
13–17

Reminder: A means *Do not use a calculator.*

Use the numbers in the following table to answer the
questions below. You may not use a number more than once.

19	85.2	533	571
88.2	525	20	17.5
400	261	20.5	125
7	23	901	30

1. Circle two numbers whose sum is 832.

2. Make an X in the boxes containing three
numbers whose sum is 57.

3. Make a check mark in the boxes containing
two prime numbers whose sum is 42.

4. Make a star in the boxes containing two numbers whose sum is 658.

5. Make a triangle in the boxes containing two numbers whose sum is 105.7.
Explain how you found the answer.

_____

_____

_____

_____

Solve Problems 6–9 using any method you want. Show your work in the space below.

6. $3{,}804 + 768 =$ _____

7. $2.83 + 1.57 =$ _____

8. $33 + 148 + 65 =$ _____

9. $1.055 + 0.863 =$ _____

Practice

10. $73 - 26 =$ _____

11. $727 - 519 =$ _____

12. $27 \div 9 =$ _____

13. $4\overline{)34} \rightarrow$ _____

Another Number Hunt

Use the numbers in the following table to answer the questions below.
You may not use a number more than once.

17	15	9	75.03
100.9	803	25	451
1,500	5,000	1	3,096
299	703	75.3	40.03

1. Circle two numbers whose difference is 152.

2. Make an X in the boxes of two numbers whose difference is 25.6.

3. Make a check mark in the boxes of two numbers whose difference is greater than 1,000.

4. Make a star in the boxes of two numbers whose difference is less than 10.

5. Make a triangle in the boxes of two numbers whose difference is equal to the sum of 538 and 259.

6. Use diagonal lines to shade the boxes of two numbers whose difference is equal to 4^2.

Subtract. Show your work for one problem on the grid below.

7. $247 - 186 =$ _____

8. _____ $= 405 - 268$

9. $24.5 - 18.7 =$ _____

10. _____ $= 62.7 - 43.85$

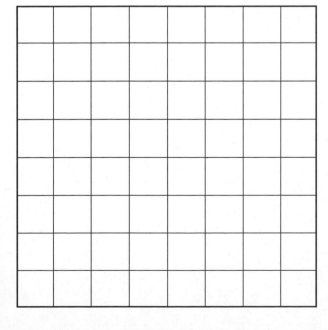

Practice

11. $48 \div 8 =$ _____

12. $81,447 + 2,571 =$ _____

13. $\$451.17 + \$2.81 =$ _____

14. $14 * 7 =$ _____

15. $98 \div 7 =$ _____

31

Name _____ Date _____ Time _____

Open Sentences and Number Stories

Read each problem. Fill in the blanks and solve the problem.

SRB
223

1. Althea and her brother collect baseball cards. Althea has 148 cards.
Her brother has 127 cards. How many cards do they have altogether?

   a. List the numbers needed to solve the problem. _____

   b. Describe what you want to find. _____

   c. Open number sentence: _____

   d. Solution: _____   **e.** Answer: _____
                                                               (unit)

2. Mark bought a hamburger for $3.89 and a drink for $1.49. If he paid with a
$20 bill, how much change did he receive?

   a. List the numbers needed to solve the problem. _____

   b. Describe what you want to find. _____

   c. Open number sentence: _____

     _____

   d. Solution: _____   **e.** Answer: _____
                                                            (unit)

3. Fran has four pieces of ribbon. Each piece of ribbon is a different length:
0.6 meters long, 1.15 meters long, 1.35 meters long, and 0.925 meters long.
How many meters of ribbon does Fran have in all?

   a. List the numbers needed to solve the problem.

     _____

   b. Describe what you want to find. _____

   c. Open number sentence:

     _____

   d. Solution: _____

   e. Answer: _____
                                   (unit)

Comparing Reaction Times

SRB
114

Use your Grab-It Gauge. Collect reaction-time data from two people at home. At least one of these people should be an adult.

1.

Person 1	
Left	**Right**

2.

Person 2	
Left	**Right**

3. Median times:

Left hand _____

Right hand _____

4. Median times:

Left hand _____

Right hand _____

5. How do the results for the two people compare to your class data?

_____

_____

_____

Practice

6. 2,683 + 2,939 = _____

7. 3,702 ∗ 8 = _____

8. 604 − 86 = _____

9. 39 ÷ 3 = _____

35

STUDY LINK 2·6

How Likely Is Rain?

Many years ago, weather reports described the chances of rain with phrases such as *very likely, unlikely,* and *extremely unlikely.* Today, the chances of rain are almost always reported as percents. For example, "There is a 50% chance of rain tonight."

1. Use the Probability Meter Poster to translate phrases into percents.

Phrase	Percent
Unlikely	30%
Very likely	
Very unlikely	
Likely	
Extremely unlikely	

2. Use the Probability Meter Poster to translate percents into phrases.

Percent	Phrase
30%	Unlikely
5%	
99%	
20%	
80%	
35%	
65%	
45%	

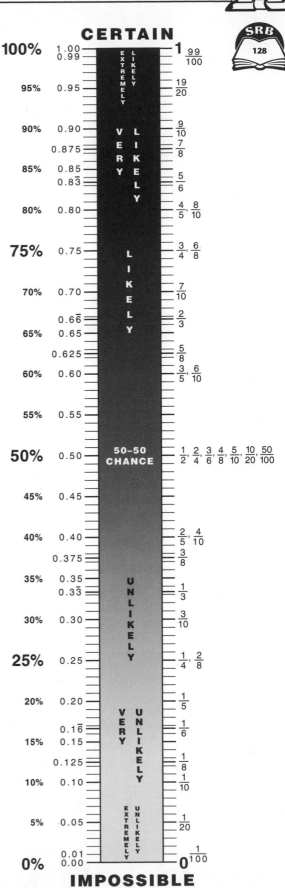

STUDY LINK 2·8 Estimating and Multiplying

◆ For each problem, make a magnitude estimate.

◆ Circle the appropriate box. Do not solve the problem.

◆ Then choose 3 problems to solve. Show your work on the grid.

SRB
247

1. 8 * 19 _____

10s	100s	1,000s	10,000s

_____

How I estimated

2. 155 * 6 _____

10s	100s	1,000s	10,000s

_____

How I estimated

3. 37 * 58 _____

10s	100s	1,000s	10,000s

_____

How I estimated

4. 5 * 4.2 _____

10s	100s	1,000s	10,000s

_____

How I estimated

5. 9.3 * 2.8 _____

10s	100s	1,000s	10,000s

_____

How I estimated

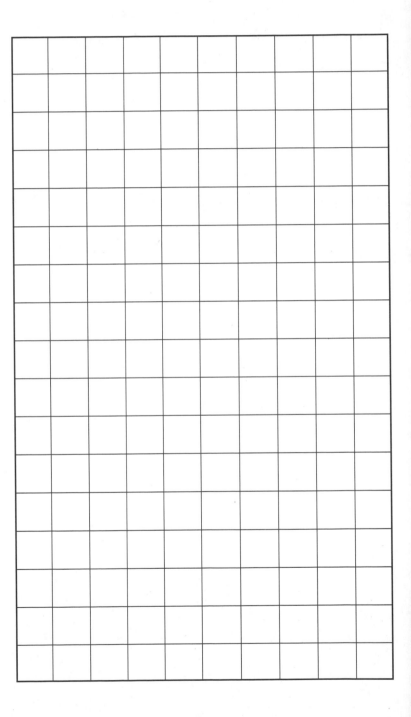

STUDY LINK
2·9

Multiply with the Lattice Method

SRB
20

For each problem:

◆ Make a magnitude estimate. Circle the appropriate box.

◆ Solve using the lattice method. Show your work in the grids.

1. 94 * 73 = _____

| 10s | 100s | 1,000s | 10,000s |

2. 24 * 3.7 = _____

| 0.1s | 1s | 10s | 100s |

3. 5.4 * 6.18 = _____

| 0.1s | 1s | 10s | 100s |

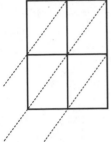

4. 384 * 261 = _____

| 100s | 1,000s | 10,000s | 100,000s |

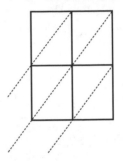

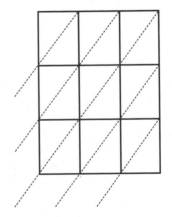

5. 17.7 * 19.3 = _____

| 0.1s | 1s | 10s | 100s |

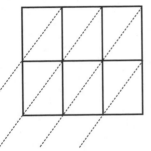

Practice

6. 7,402 + 2,587 = _____

7. 37 ÷ 7 → _____

8. 328 − 237 = _____

9. $15.75 + $3.25 = _____

43

STUDY LINK 2·10 Place-Value Puzzles

Millions			Thousands			Ones		
Hundred-millions	Ten-millions	Millions	Hundred-thousands	Ten-thousands	Thousands	Hundreds	Tens	Ones

Use the clues to solve the puzzles.

Puzzle 1

○ The value of the digit in the **thousandths** place is equal to the sum of the measures of the angles in a triangle (180°) divided by 30.

○ If you multiply the digit in the **tens** place by 1,000; the answer will be 9,000.

◇ Double 35. Divide the result by 10. Write the answer in the **tenths** place.

○ The **hundreds**-place digit is $\frac{1}{2}$ the value of the digit in the thousandths place.

○ When you multiply the digit in the **ones** place by itself, the answer is 0.

○ Write a digit in the **hundredths** place so that the sum of all six digits in this number is 30.

What is the number? _____ _____ _____ . _____ _____ _____

Puzzle 2

○ Double 12. Divide the result by 8. Write the answer in the **thousands** place.

○ If you multiply the digit in the **hundredths** place by 10, your answer will be 40.

◇ The **tens**-place digit is a prime number. If you multiply it by itself, the answer is 49.

○ Multiply 7 and 3. Subtract 12. Write the answer in the **thousandths** place.

◇ Multiply the digit in the hundredths place by the digit in the thousands place. Subtract 7 from the result. Write the digit in the **tenths** place.

○ The digit in the **ones** place is an odd digit that has not been used yet.

○ The value of the digit in the **hundreds** place is the same as the number of sides of a quadrilateral.

What is the number? _____ , _____ _____ _____ . _____ _____ _____

Check: The sum of the answers to both puzzles is 3,862.305.

Practice

3. 7,772 + 1,568 = _____

4. 472 − 228 = _____

5. 826 * 54 = _____

6. 59 / 3 → _____

45

Unit 3: Family Letter

Geometry Explorations and the American Tour

In Unit 3, your child will set out on the American Tour, a yearlong series of mathematical activities examining historical, demographic, and environmental features of the United States. The American Tour activities will develop your child's ability to read, interpret, critically examine, and use mathematical information presented in text, tables, and graphics. These math skills are vital in our technological age.

Many American Tour activities rely on materials in the American Tour section of the *Student Reference Book.* This section—part historical atlas and part almanac—contains maps, data, and other information from a wide range of sources: the U.S. Census Bureau, the National Weather Service, and the National Geographic Society.

Unit 3 also will review some geometry concepts from earlier grades while introducing and expanding on others. In *Fourth Grade Everyday Mathematics,* students used a compass to construct basic shapes and create geometric designs. In this unit, your child will extend these skills and explore concepts of congruent figures (same size, same shape), using a compass and straightedge. In addition, students will use another tool, the Geometry Template. It contains protractors and rulers for measuring, as well as cutouts for drawing a variety of geometric figures.

Finally, students will explore the mathematics and art of tessellations—patterns of shapes that cover a surface without gaps or overlaps. They will use math tools to create their own designs.

You can help your child by asking questions about information presented in newspaper and magazine tables and graphics. Also, the world is filled with many 2-dimensional and 3-dimensional geometric forms: angles, line segments, curves, cubes, cylinders, spheres, pyramids, and so on. Many wonderful geometric patterns can be seen in nature as well as in the things that people create. It will be helpful for you and your child to look for and talk about geometric shapes throughout the year.

Please keep this Family Letter for reference as your child works through Unit 3.

Vocabulary

Important terms in Unit 3:

acute angle An angle with a measure greater than 0 degrees and less than 90 degrees.

Acute angle

adjacent angles Two angles with a common side and vertex that do not otherwise overlap. In the diagram, angles 1 and 2 are adjacent angles. Angles 2 and 3, angles 3 and 4, and angles 4 and 1 are also adjacent.

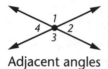

Adjacent angles

congruent Having exactly the same shape and size.

Congruent triangles

diameter A line segment that passes through the center of a circle (or sphere) and has endpoints on the circle (or sphere); also, the length of this line segment. The diameter of a circle or sphere is twice the length of its radius.

equilateral triangle A triangle with all three sides the same length. In an equilateral triangle, all three angles have the same measure.

Equilateral triangles

obtuse angle An angle with a measure greater than 90 degrees and less than 180 degrees.

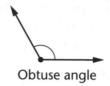

Obtuse angle

radius A line segment from the center of a circle (or sphere) to any point on the circle (or sphere); also, the length of this line segment.

right angle An angle with a measure of 90 degrees.

Right angle

tessellation An arrangement of shapes that covers a surface completely without overlaps or gaps. Also called *tiling*.

A tessellation

vertical (opposite) angles The angles made by intersecting lines that do not share a common side. Vertical angles have equal measures. In the diagram, angles 2 and 4 are a pair of vertical angles. Angles 1 and 3 are another pair of vertical angles.

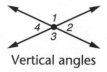

Vertical angles

48

Building Skills through Games

In Unit 3, your child will practice geometry and computation skills by playing the following games. For detailed instructions, see the *Student Reference Book.*

Angle Tangle See *Student Reference Book,* page 296
Two players will need a protractor and a straightedge to play this game. Playing *Angle Tangle* gives students practice in drawing and measuring angles.

High-Number Toss: Decimal Version See *Student Reference Book,* page 321
This game practices concepts of place value and standard notation. It requires 2 players and number cards 0–9 (4 of each).

Multiplication Top-It See *Student Reference Book,* page 334
This game practices the basic multiplication facts. It requires a deck of cards with 4 each of the numbers 1–10, and can be played by 2–4 players.

Polygon Capture See *Student Reference Book,* page 328
This game uses 16 polygons and 16 Property Cards, and is played by partners or 2 teams each with 2 players. *Polygon Capture* practices identifying properties of polygons related to sides and angles.

Do-Anytime Activities

To work with your child on the concepts taught in this unit and in previous units, try these interesting and rewarding activities:

1. Together, read the book *A Cloak for the Dreamer* by Marilyn Burns.

2. When you are at home or at a store, ask your child to identify different types of polygons such as triangles, squares, pentagons, and hexagons.

3. Visit the Web site for the U.S. Bureau of the Census at http://www.census.gov/. Have your child write three interesting pieces of information that he or she learned from the Web site.

4. Look for examples of bar graphs in newspapers or magazines. Ask your child to explain the information shown by a graph.

As You Help Your Child with Homework

As your child brings assignments home, you may want to go over the instructions together, clarifying them as necessary. The answers listed below will guide you through this unit's Study Links.

Study Link 3·1

1. Illinois

2. 851,000; 4,822,000; 8,712,000; 12,051,000

3. 3,971,000   4. 3,890,000   5. 3,339,000

6. The population increases by about 4,000,000 every fifty years.

7. About 16,000,000   8. About 14,000,000

Study Link 3·2

1. A   2. 5,472,000   3. H

4. a. About 250,000,000   b. About 55%

Study Link 3·3

1. 60°; 90°; 60°   2. 120°; 60°; 60°

3. 90°; 135°; 135°   4. 30°; 75°

Study Link 3·4

1. 70°   2. 50°   3. 110°   4. 130°

5. 60°   6. 180°   7. 120°   8. 90°

9. 50°   10. 150°   11. 170°

Study Link 3·5

1. acute; 12°   2. acute; 65°   3. obtuse; 103°

4. Sample answer: Angle *D* and angle *E*

5. Sample answer: Angle *D* and angle *F*

6. Sample answer: Angle *G* and angle *H*

9. 14,670   11. 11R1

Study Link 3·6

1. scalene   2. isosceles   3. isosceles; right

4. equilateral; isosceles

5. Objects and types of angles vary.

6. 11,761   7. 5,750   8. 42,405   9. 11

Study Link 3·7

Sample answers are given for Problems 1–5.

1. The pentagon is the only shape that is not regular.

2. The oval is the only shape that is curved.

3.  The crossed-out shape is the only shape that is not convex.

4. The trapezoid is the only shape without two pairs of parallel sides.

Study Link 3·8

1.–3. Samples of tessellations vary.

Study Link 3·9

1. Sample answer: Draw a line between two of the vertices to create two triangles. Since the sum of the angles in each triangle is 180°, the sum of the angles in a quadrangle is 360°.

2. 360°

3. a.–b.   c.–d.

Study Link 3·10

1. Sample answers are given.

a.   b.

c.   d.

2.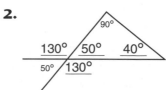

3. a. 2   b. 70°   c. 360°   d. trapezoid

**STUDY LINK
3·1**

Population Data

State	1850	1900	1950	2000
Ohio	1,980,000	4,158,000	7,947,000	11,319,000
Indiana	988,000	2,516,000	3,934,000	6,045,000
Illinois	851,000	4,822,000	8,712,000	12,051,000
Michigan	398,000	2,421,000	6,372,000	9,679,000
Wisconsin	305,000	2,069,000	3,435,000	5,326,000
Minnesota	6,000	1,751,000	2,982,000	4,830,000
Iowa	192,000	2,232,000	2,621,000	2,900,000
Missouri	682,000	3,107,000	3,955,000	5,540,000

1. Which state had the largest population growth from 1850 to 2000? _____

2. Record the population figures for this state below the timeline.

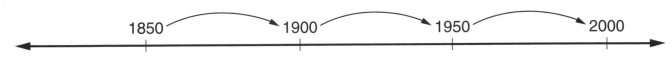

1850   1900   1950   2000

_____ _____ _____ _____

Find the increases for this state for each of the following time spans:

3. 1850–1900 _____ **4.** 1900–1950 _____

5. 1950–2000 _____

6. Are these increases similar or different? Explain.

_____

_____

Estimate the state's population:

7. In 2050 _____ **8.** In 2025 _____

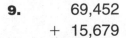
9.   69,452
      + 15,679

10.   178
      − 139

11.   43
       * 14

12. 58 ÷ 7→ _____

51

An Unofficial Census

STUDY LINK
3·2

In 1991, author Tom Heymann took an unofficial U.S. census. The table shows
how many people believed various common sayings, based on the sample of the
population that he surveyed.

	Saying	Number Who Believe Saying Is True
A	Look before you leap.	175,104,000
B	The grass is always greener on the other side of the fence.	69,312,000
C	Haste makes waste.	153,216,000
D	Beauty is only skin deep.	149,568,000
E	Don't cry over spilled milk.	160,512,000
F	The early bird catches the worm.	136,800,000
G	A penny saved is a penny earned.	155,040,000
H	Don't count your chickens before they hatch.	169,632,000

Source: *The Unofficial U.S. Census,* by Tom Heymann. Ballantine Books, 1991

1. Which saying had the largest number of believers? _____

2. How many more people believed saying E than saying G? _____

3. Which saying had about 100 million more believers than saying B? _____

4. a. About $\frac{7}{10}$ of the U.S. population in 1991 believed
saying A to be true. What was the total population? _____

b. About what percent of the total
population believed saying F to be true? _____

Practice		

5. 256
       − 148

6. 26,551
      + 2,558

7.   36
       * 27

8. 54 ÷ 3 = _____

9. 74 ÷ 8 → _____

Finding Angle Measures

Figure out the angle measures for the labeled angles in the patterns below.
Remember that there are 360° in a circle and 180° in a straight line. Use the
Geometry Template, or cut out the shapes at the bottom of this page to help you.
Do not use a protractor.

1.

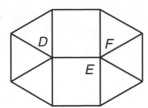

m∠D = _____

m∠E = _____

m∠F = _____

2.

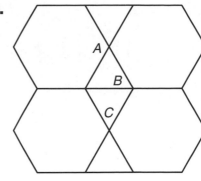

m∠A = _____

m∠B = _____

m∠C = _____

3.

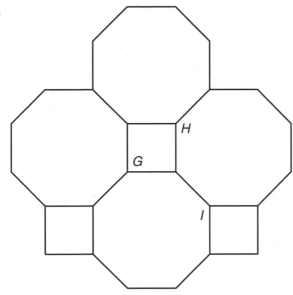

m∠G = _____

m∠H = _____

m∠I = _____

4.

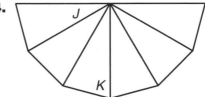

m∠J = _____

m∠K = _____

5. On the back of this page, explain how
you found the measure of ∠I.

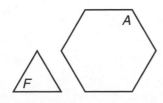

STUDY LINK
3·4

Angle Measures

SRB
204–205

Find the approximate measure of each angle at the right.

1. measure of ∠CAT = _____

2. m∠BAR = _____

3. m∠RAT = _____

4. m∠CAB = _____

5. m∠BAT = _____

6. m∠CAR = _____

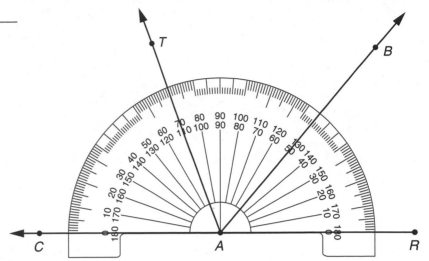

Find the approximate measure of each angle at the right.

7. m∠MEN = _____

8. m∠DEN = _____

9. m∠MET = _____

10. m∠MED = _____

11. m∠TEN = _____

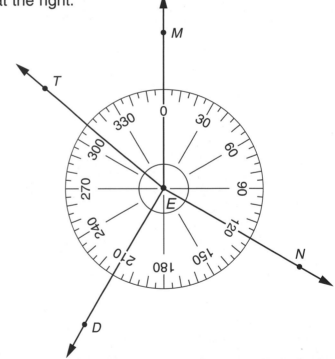

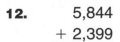
Practice

12. 5,844
      + 2,399

13. 238
      − 129

14. 234
      * 22

15. 60 ÷ 5 = _____

16. 50 ÷ 6 → _____

STUDY LINK
3·5
Angles in Figures

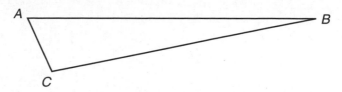

Circle *acute, right,* or *obtuse* for each angle in triangle *ABC*.
Then measure each angle.

1. ∠*ABC*   acute   right   obtuse   m∠*ABC* = _____

2. ∠*CAB*   acute   right   obtuse   m∠*CAB* = _____

3. ∠*BCA*   acute   right   obtuse   m∠*BCA* = _____

Use the figure at the right to do Problems 4–6.

4. Name a pair of adjacent angles.

_____ and _____

5. Name a pair of vertical angles.

_____ and _____

6. Name a pair of opposite angles.

_____ and _____

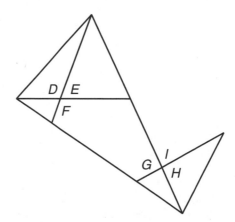

| **Practice** |

7.   7,568
     + 9,217
     ‾‾‾‾‾‾‾

8.   415
     − 207
     ‾‾‾‾‾

9.   326
     * 45
     ‾‾‾‾

10. 68 ÷ 4 = _____

11. 78 ÷ 7 → _____

STUDY LINK
3·6

Triangle and Angle Review

SRB
144

For each triangle below, fill in the ovals for all the names that apply.

1.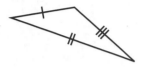

2.

3.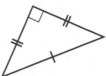

4.

◯ equilateral	◯ equilateral	◯ equilateral	◯ equilateral
◯ isosceles	◯ isosceles	◯ isosceles	◯ isosceles
◯ right	◯ right	◯ right	◯ right
◯ scalene	◯ scalene	◯ scalene	◯ scalene

On the back of this page, draw three angles of different sizes that you find at home. (For example, you could trace one corner of a book.) For each angle, name the object that has the angle. Then use words from the Word Bank to name each angle.

5. a. Object _____

Type of angle _____

b. Object _____

Type of angle _____

c. Object _____

Type of angle _____

Word Bank		
acute	obtuse	right
adjacent	reflex	straight

Practice

6. $4{,}117 + 3{,}682 + 3{,}962 =$ _____

7. $8{,}036 - 2{,}286 =$ _____

8. $8{,}481 * 5 =$ _____

9. $99 ÷ 9 =$ _____

Name _____ Date _____ Time _____

Odd Shape Out

In each set of shapes, there is one shape that doesn't belong. Cross out that shape and tell why it doesn't belong. (There may be more than one possible reason. What's important is having a good reason for crossing out a shape.)

1.

Reason: _____

_____

2.

Reason: _____

_____

3.

Reason: _____

_____

4.

Reason: _____

_____

5. Make up your own "Odd Shape Out" problem on the back of this page.

Practice

6. $1,042 + 2,834 + 4,096 =$ _____

7. $9,062 - 3,718 =$ _____

8. $9,109 * 9 =$ _____

9. $58 \div 6 \rightarrow$ _____

STUDY LINK
3·8

Tessellation Museum

A **tessellation** is an arrangement of repeated, closed shapes that completely covers a surface, without overlaps or gaps. Sometimes only one shape is used in a tessellation. Sometimes two or more shapes are used.

SRB
160 161

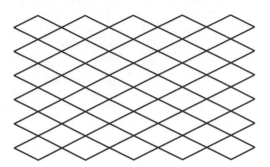
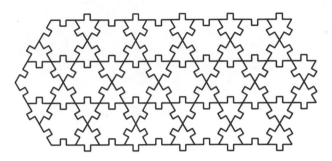

1. Collect tessellations. Look in newspapers and magazines. Ask people at home to help you find examples.

2. Ask an adult whether you may cut out the tessellations. Tape your tessellations onto this page in the space below.

3. If you can't find tessellations in newspapers or magazines, look around your home at furniture, wallpaper, tablecloths, or clothing. In the space below, sketch the tessellations you find.

Practice

🚫🧮

4. 1,987 + 6,213 + 2,046 = _____

5. 4,615 − 3,148 = _____

6. 3,714 * 8 = _____

7. 39 / 7 → _____

65

STUDY LINK 3·9 | **Sums of Angle Measures**

1. Describe one way to find the sum of the angles in a quadrangle without using a protractor. You might want to use the quadrangle at the right to illustrate your explanation.

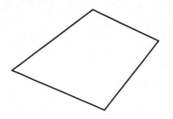

_____

_____

_____

_____

_____

2. The sum of the angles in a quadrangle is _____.

3. Follow these steps to check your answer to Problem 2.

   a. With a straightedge, draw a large quadrangle on a separate sheet of paper.

   b. Draw an arc in each angle.

   c. Cut out the quadrangle and tear off part of each angle.

   d. Tape or glue the angles onto the back of this page so that the angles touch but do not overlap.

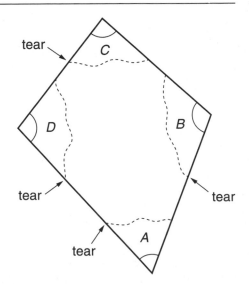

Practice

4. 3,007 + 1,251 + 980 = _____

5. 4,310 − 1,290 = _____

6. 3,692 * 6 = _____

7. 67 ÷ 8 → _____

STUDY LINK 3·10 | **Polygons and Their Measures**

SRB
142 143

1. Draw each of the following figures.

   a. a polygon

   b. a triangle with no equal sides

   c. a quadrangle with one right angle

   d. a quadrangle with no pairs of parallel sides

2. Without using a protractor, record the missing angle measurements in the figure to the right.

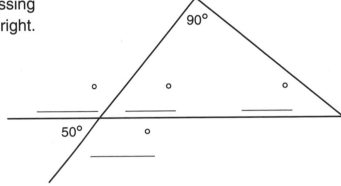

3. Use the figure to the right to answer the questions.

   a. How long is line segment *CD*? _____ cm

   b. What is the measure of angle *A*? _____

   c. What is the sum of the measures of all

   the angles? _____

   d. What is a geometric name for the figure? _____

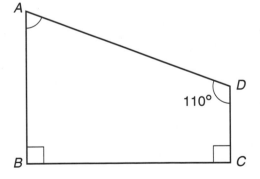

Practice

4. 1,476 + 2,724 + 3,241 = _____

5. 4,002 − 1,361 = _____

6. 5,031 * 4 = _____

7. 27 ÷ 9 = _____

69

STUDY LINK 3·11 | Unit 4: Family Letter

Division

Unit 4 begins with a review of division facts and the relationship between division and multiplication. Emphasis is on fact families. A person who knows that $4 * 5 = 20$ also knows the related facts $5 * 4 = 20$, $20 \div 4 = 5$, and $20 \div 5 = 4$.

We will develop strategies for dividing mentally. Challenge your child to a game of *Division Dash* to help him or her practice. You'll find the rules in the *Student Reference Book,* page 303.

These notations for division are equivalent:	
$12\overline{)246}$	$246 \div 12$
$246 / 12$	$\dfrac{246}{12}$

In *Fourth Grade Everyday Mathematics,* students were introduced to a method of long division called the partial-quotients division algorithm. This algorithm is easier to learn and apply than the traditional long-division method. It relies on "easy" multiplication, and it can be quickly employed by students who struggle with traditional computation.

In this method, a series of partial answers (partial quotients) are obtained, and then added to get the final answer (the quotient). After your child has worked with this method, you might ask him or her to explain the example below:

$$
\begin{array}{r|l}
12\overline{)158} & \\
-120 & 10 \\
\hline
38 & \\
-36 & 3 \\
\hline
2 & 13
\end{array}
$$

    ↑   ↑

Remainder Quotient

In the coming unit, we will review the partial-quotients algorithm and extend it to decimals.

Your child will practice using this division algorithm, as well as others, if he or she chooses. The partial-quotients division algorithm and another method called column division are described in the *Student Reference Book.*

When we solve division number stories, special attention will be placed on interpreting the remainder in division.

The American Tour will continue as the class measures distances on maps and uses map scales to convert the map distances to real-world distances between cities, lengths of rivers, and so on.

Please keep this Family Letter for reference as your child works through Unit 4.

Vocabulary

Important terms in Unit 4:

dividend In division, the number that is being divided. For example, in 35 ÷ 5 = 7, the dividend is 35.

divisor In division, the number that divides another number. For example, in 35 ÷ 5 = 7, the divisor is 5.

map legend (map key) A diagram that explains the symbols, markings, and colors on a map.

map scale The ratio of a distance on a map, globe, or drawing to an actual distance.

number sentence Two expressions with a relation symbol (=, <, >, ≠, ≤, or ≥). For example, 5 + 5 = 10 and 6 * (43 + 7) = 300 are number sentences. Compare to *open sentence.*

open sentence A *number sentence* with one or more *variables*. For example, *x* + 3 = 5 is an open sentence.

quotient The result of dividing one number by another number. For example, in 35 ÷ 5 = 7, the quotient is 7.

remainder The amount left over when one number is divided by another number. For example, if 38 books are divided into 5 equal piles, there are 7 books per pile, with 3 books remaining. In symbols, 38 ÷ 5 → 7 R3.

variable A letter or other symbol that represents a number. A variable can represent one specific number. For example, in the number sentence 5 + *n* = 9, only *n* = 4 makes the sentence true. A variable may also stand for many different numbers. For example, *x* + 2 < 10 is true if *x* is any number less than 8.

Do-Anytime Activities

To work with your child on the concepts taught in this unit and in previous units, try these interesting and rewarding activities:

1. Provide your child with opportunities to look at maps from various parts of the country. Ask him or her to explain the map legend and map scale, and to find the distances between two cities or places of interest.

2. Read the book *A Remainder of One,* by Elinor J. Pinczes.

3. Play *Division Dash, First to 100, Divisibility Dash, Division Top-It* or *Name that Number* as described in the *Student Reference Book.*

4. Ask your child to write number stories that can be solved using division. Help your child solve those problems, and then identify how the quotient and remainder are used to answer the question in the number story.

Building Skills through Games

In Unit 4, your child will practice division as well as other skills by playing these and other games. For detailed instructions, see the *Student Reference Book.*

Divisibility Dash See *Student Reference Book,* page 302
This is a game for two to three players and requires a set of number cards. Playing *Divisibility Dash* provides practice recognizing multiples and using divisibility rules in a context that also develops speed.

Division Dash See *Student Reference Book,* page 303
This is a game for one or two players. Each player will need a calculator. Playing *Division Dash* helps students practice division and mental calculation.

Division Top-It See *Student Reference Book,* page 334
This is a game for two to four players and requires number cards. Playing Division Top-It provides practice recognizing multiples and applying division facts and extended facts.

First to 100 See *Student Reference Book,* page 308
This is a game for two to four players and requires 32 Problem Cards and a pair of six-sided dice. Players answer questions after substituting numbers for the variable on Problem Cards. The questions offer practice on a variety of mathematical topics.

Name That Number See *Student Reference Book,* page 325
This is a game for two or three players using the Everything Math Deck or a complete deck of number cards. This game provides a review of operations with whole numbers.

As You Help Your Child with Homework

As your child brings assignments home, you may want to go over the instructions together, clarifying them as necessary. The answers listed below will guide you through this unit's Study Links.

Study Link 4·1

1. 19; Sample answer: 30 and 27

2. 12; Sample answer: 80 and 16

3. 2,000 mi   **4.** 5 lb

5. 878; 1,803 − 878 = 925; 925 + 878 = 1,803; 878 + 925 = 1,803

6. 875; 377 + 498 = 875; 875 − 377 = 67; 875 − 498 = 67

Study Link 4·2

1. 10, 10, 10, and 3   **2.** 27 R4   **3.** 42 R4

4. 32 R5   **5.** 24

6. 3,985; 3,985 − 168, or 3,817 = 3,817, or 168

7. 52,236; 281, or 52,236 + 52,236 or 281 = 52,517

Study Link 4·3

1. a. About 1 mi   **b.** About $1\frac{1}{2}$ mi

2. a. About $3\frac{3}{4}$ in.   **b.** About $1\frac{7}{8}$ mi

3. 188; 188 + 188 = 376

4. 4,148; 4,148 − 3,997, or 151 = 151, or 3,997

Study Link 4·4

1. 71   **2.** 53   **3.** 82 R22

4. 26 R10 **5.** 83 pages

6. 2,814; 2,814 − 814, or 68 = 68, or 814

7. 3,296; 165; 3,296 + 3,296; 165 = 3,461

Study Link 4·5

Estimates vary. Sample estimates are given for Problems 1–6.

1. The 10s box should be circled; 60 ÷ 6 = 10; 13.1

2. The 100s box should be circled; 300 ÷ 3 = 100; 129

3. The 1s box should be circled; 30 ÷ 10 = 3; $3.69

4. The 10s box should be circled; 800 ÷ 40 = 20; 23

5. The 100s box should be circled; 1,000 ÷ 5 = 200; 169

6. The 1s box should be circled; 18 ÷ 9 = 2; 1.76

7. 14.544; 14.544 − 8.54, or 6.004 = 6.004, or 8.54

Study Link 4·6

1. $6.25; Reported it as a fraction or decimal; Sample answer: The cost per game is exact, so the answer needs to be exact.

2. 7; Ignored it; Sample answer: The remaining $4.00 is not enough to buy another pizza, and is ignored.

3. 15 R1   **4.** 52,836

Study Link 4·7

1. 49   **2.** 780   **3.** 610

Answers vary for Problems 4–11.

12. 3,985   **13.** 52,236

STUDY LINK 4·1 | Uses of Division

Use multiplication and division facts to solve the following problems mentally.
Remember: Break the number into two or more friendly parts.

Example: How many 4s in 71?

Break 71 into smaller, friendly numbers. Here are two ways.

◆ 40 and 31. Ask yourself: *How many 4s in 40?* (10) *How many 4s in 31?* (7 and 3 left over) Think: *What multiplication fact for 4 has a product near 31?* (4 ∗ 7 = 28) Total = 17 and 3 left over.

◆ 20, 20, 20, and 11. Ask yourself: *How many 4s in 20?* (5) *How many 4s in three 20s?* (15) *How many 4s in 11?* (2 and 3 left over) Total = 17 and 3 left over.

So 71 divided by 4 equals 17 with 3 left over.

1. 57 divided by 3 equals _____.

_____
(friendly parts for 57)

2. 96 divided by 8 equals _____.

_____
(friendly parts for 96)

3. The diameter of Earth, about 8,000 miles, is about 4 times the diameter of the moon. What is the approximate diameter of the moon?

8,000 mi

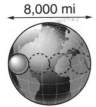

4. The weight of an object on Earth is 6 times heavier than its weight on the moon. An object that weighs 30 lb on Earth weighs how many pounds on the moon?

_____
unit

_____
unit

Practice

Solve. Then write the other problems in the fact families.

5. 1,803 − 925 = _____

_____

_____

6. 498 + 377 = _____

_____

_____

STUDY LINK 4·2 Division

Here is the partial-quotients algorithm using a friendly numbers strategy.

$7\overline{)237}$

Rename dividend (use multiples of the divisor):
237 = 210 + 21 + 6

$\begin{array}{r} -210 \\ \hline 27 \end{array}$ 30

How many 7s are in 210? 30
The first partial quotient. 30 * 7 = 210
Subtract. 27 is left to divide.

$\begin{array}{r} -21 \\ \hline \end{array}$ 3

How many 7s are in 27? 3
The second partial quotient. 3 * 7 = 21
Subtract. 6 is left to divide.

6   33

Add the partial quotients: 30 + 3 = 33

↑   ↑
Remainder Quotient   Answer: 33 R6

1. Another way to rename 237 with multiples of 7 is

237 = 70 + 70 + 70 + 21 + 6

If the example had used this name for 237, what would the partial quotients have been?

_____

2. $6\overline{)166}$

Answer: _____

3. 214 / 5

Answer: _____

4. 485 ÷ 15

Answer: _____

5. $17\overline{)408}$

Answer: _____

Practice

6. 3,817 + 168 = _____

Check: _____ − _____ = _____

7. 52,517 − 281 = _____

Check: _____ + _____ = _____

STUDY LINK 4·3 | **Distance to School**

There are two ways to go from Josephina's house to school. She can take Elm Street and then Washington Avenue. She can also take Snakey Lane.

Use the map and scale below to answer the questions.

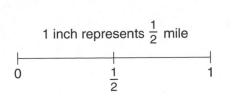

1 inch represents $\frac{1}{2}$ mile

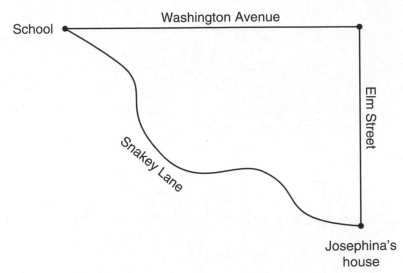

1. Josephina started walking from home to school along Elm Street.

   a. How far would Josephina walk before she
   turned onto Washington Avenue?   _____

   b. How far would she be from school when she
   turned the corner?   _____

2. Suppose Josephina could take a straight path from her house to school.
   Estimate the distance.

   a. Draw and measure a straight line on the map
   from Josephina's house to the school.   _____

   b. Use the scale to measure this distance
   in miles.   _____

Practice

3. 376 − 188 = _____

   Check: _____ + _____ = _____

4. 3,997 + 151 = _____

   Check: _____ − _____ = _____

STUDY LINK 4·4   Division

Here is an example of the partial-quotients algorithm using an "at least...not more than" strategy.

$$8)\overline{185}$$ Begin estimating with multiples of 10.

$-\ 80$ **10**	How many 8s are in 185? At least 10. The first partial quotient. $10 * 8 = 80$
$\overline{105}$	Subtract. 105 is left to divide.
$-\ 80$ **10**	How many 8s are in 105? At least 10. The second partial quotient. $10 * 8 = 80$
$\overline{25}$	Subtract. 25 is left to divide.
$-\ 24$ **3**	How many 8s are in 25? At least 3. The third partial quotient. $3 * 8 = 24$
$\overline{1}$	Subtract. 1 is left to divide.
1 **23**	Add the partial quotients: $10 + 10 + 3 = 23$
↑ ↑	

Remainder Quotient   Answer: 23 R1

Solve.

1. $639 \div 9$

Answer: _____

2. $954 \div 18$

Answer: _____

3. $1,990 / 24$

Answer: _____

4. $972 / 37$

Answer: _____

5. Robert is making a photo album. 6 photos fit on a page. How many pages will he need for 497 photos? _____ pages

---

Practice

6. $2,746 + 68 =$ _____

Check: _____ − _____ = _____

7. $3,461 − 165 =$ _____

Check: _____ + _____ = _____

STUDY LINK 4·5 Estimate and Calculate Quotients

For each problem:

◆ Make a magnitude estimate of the quotient. Ask yourself:
   Is the answer in the tenths, ones, tens, or hundreds?

◆ Circle a box to show the magnitude of your estimate.

◆ Write a number sentence to show how you estimated.

◆ If there is a decimal point, ignore it. Divide the numbers.

◆ Use your magnitude estimate to place the decimal point in the final answer.

◆ Check that your final answer is reasonable.

1. 6)78.6

0.1s	1s	10s	100s

How I estimated: _____

Answer: _____

2. 3)387

0.1s	1s	10s	100s

How I estimated: _____

Answer: _____

3. $29.52 ÷ 8

0.1s	1s	10s	100s

How I estimated: _____

Answer: _____

4. 989 ÷ 43

0.1s	1s	10s	100s

How I estimated: _____

Answer: _____

5. 845 / 5

0.1s	1s	10s	100s

How I estimated: _____

Answer: _____

6. 15.84 / 9

0.1s	1s	10s	100s

How I estimated: _____

Answer: _____

Practice

7. 8.54 + 6.004 = _____

Check: _____ − _____ = _____

STUDY LINK 4·6

Division Number Stories with Remainders

For each number story draw a picture or write a number sentence on the back of this page. Then divide to solve the problem. Decide what to do about the remainder. Explain what you did.

Example:

You need to set up benches for a picnic. Each bench seats 7 people. You expect 25 people to attend. How many benches do you need?

$$25 \div 7 = b$$

Circle what you did with the remainder.

How many benches?
7 seats per bench

| 7 |
| 7 |
| 7 |
| 4 |

} 25 people

_4_ benches

Ignored it   Reported it as a fraction or decimal   ⟨Rounded the answer up⟩

Why? _3 benches seat 21 people. One more bench is needed._

1. It costs $50.00 to be a member of a soccer team. The team plays 8 games during the season. What is the cost per game? $ _____

   Circle what you did with the remainder.

   Ignored it   Reported it as a fraction or decimal   Rounded the answer up

   Why? _____

   _____

2. Lynn is having a party. Pizzas cost $8.00 each. How many pizzas can she buy with $60.00?   _____ pizzas

   Circle what you did with the remainder.

   Ignored it   Reported it as a fraction or decimal   Rounded the answer up

   Why? _____

   _____

Practice

3. $31 \div 2 \rightarrow$ _____   4. $629 * 84 =$ _____

STUDY LINK 4·7 Variables

For Problems 1–3:

◆ Find the value of x in the first number sentence.

◆ Use this value to complete the second number sentence.

1. x = number of days in a week

$x^2 =$ _____

2. $x = \frac{1}{10}$ of 100

$x * 78 =$ _____

3. x = largest sum possible with 2 six-sided dice

$598 + x =$ _____

4. Count the number of letters in your first name and in your last name.

a. My first name has _____ letters.

b. My last name has _____ letters.

c. Find the product of these 2 numbers. Product = _____

Answer the questions in Problems 5–11 by replacing x with the product you found in Problem 4.

5. Is x a prime or a composite number? _____

6. Is $\frac{x}{30}$ less than 1? _____

7. Which is larger, $3 * x$, or $x + 100$? _____

8. What is the median and the range for this set of 3 weights: 30 pounds, 52 pounds, x pounds? _____

9. There are 200 students at Henry Clissold School. x% speak Spanish. How many students speak Spanish? _____

10. $(3x + 5) - 7 =$ _____

11. True or false: $x^2 > 30 * x$ _____

Practice

12. $3,817 + 168 =$ _____

13. $52,517 - 281 =$ _____

Unit 5: Family Letter

Fractions, Decimals, and Percents

Unit 5 focuses on naming numbers as fractions, decimals, and percents. Your child will use pattern blocks to review basic fraction and mixed-number concepts as well as notations. Your child will also formulate rules for finding equivalent fractions.

In *Fourth Grade Everyday Mathematics,* your child learned to convert easy fractions, such as $\frac{1}{2}$, $\frac{1}{4}$, $\frac{1}{10}$, and $\frac{3}{4}$, to equivalent decimals and percents. For example, $\frac{1}{2}$ can be renamed as 0.5 or 50%. Your child will now learn (with the use of a calculator) how to rename any fraction as a decimal and as a percent.

Unit 5 also introduces two new games: *Estimation Squeeze,* to practice estimating products; and *Frac-Tac-Toe,* to practice converting fractions to decimals and percents. These games, like others introduced earlier, are used to reinforce arithmetic skills. Both games use simple materials (calculator, number cards, and pennies or other counters) so you can play them at home.

Your child will study data about the past and compare it with current information as the American Tour continues.

Please keep this Family Letter for reference as your child works through Unit 5.

Vocabulary

Important terms in Unit 5:

bar graph A graph that uses horizontal or vertical bars to represent data.

circle graph A graph in which a circle and its interior are divided through its center into parts to show the parts of a set of data. The whole circle represents the whole set of data.

denominator The number below the line in a fraction. In a fraction representing a whole, or ONE, divided into equal parts, the denominator is the total number of equal parts. In the fraction $\frac{a}{b}$, b is the denominator.

equivalent fractions Fractions that have different denominators but name the same amount. For example, $\frac{1}{2}$ and $\frac{4}{8}$ are equivalent fractions.

improper fraction A fraction whose numerator is greater than or equal to its denominator. For example, $\frac{4}{3}$, $\frac{5}{2}$, $\frac{4}{4}$, and $\frac{24}{12}$ are improper fractions. In *Everyday Mathematics,* improper fractions are sometimes called "top-heavy" fractions.

mixed number A number that is written using both a whole number and a fraction. For example, $2\frac{1}{4}$ is a mixed number equal to $2 + \frac{1}{4}$.

numerator The number above the line in a fraction. In a fraction representing a whole, or ONE, divided into equal parts, the numerator is the number of equal parts that are being considered. In the fraction $\frac{a}{b}$, a is the numerator.

percent (%) Per hundred, or out of a hundred. For example, *48% of the students in the school are boys* means that, on average, 48 out of every 100 students in the school are boys.

Percent Circle A tool on the Geometry Template that is used to measure or draw figures that involve percents, such as *circle graphs*.

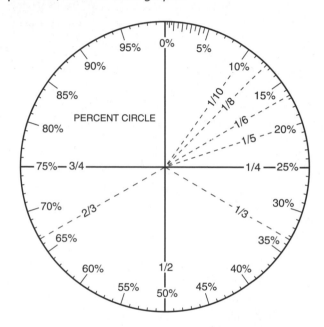

repeating decimal A decimal in which one digit or a group of digits is repeated without end. For example, 0.333... and $0.\overline{147}$ are repeating decimals.

Do-Anytime Activities

To work with your child on the concepts taught in this unit and in previous units, try these interesting and rewarding activities.

1. Help your child find fractions, decimals, and percents in the everyday world—in newspaper advertisements, on measuring tools, in recipes, in the sports section of the newspaper, and so on.

2. Over a period of time, have your child record daily temperatures in the morning and in the evening. Keep track of the temperatures in a chart. Then have your child make a graph from the data. Ask questions about the data. For example, have your child find the differences in temperatures from morning to evening or from one day to the next.

3. Practice using percents in the context of tips. For example, have your child calculate $\frac{1}{10}$ or 10% of amounts of money. Invite your child to find the tip the next time the family goes out for dinner.

4. Ask your child to identify 2-dimensional and 3-dimensional shapes around the house.

Building Skills through Games

In Unit 5, your child will practice operations and computation skills by playing the following games. For detailed instructions, see the *Student Reference Book.*

Estimation Squeeze See *Student Reference Book,* page 304.
This is a game for two players who use a single calculator. The game provides practice in estimating products.

Frac-Tac-Toe See *Student Reference Book,* pages 309–311.
This is a game for two players. Game materials include 4 each of the number cards 0–10, pennies or counters of two colors, a calculator, and a gameboard. The gameboard is a 5-by-5 number grid that resembles a bingo card. Several versions of the gameboard are shown in the *Student Reference Book.* *Frac-Tac-Toe* helps students practice converting fractions to decimals and percents.

Fraction Of See *Student Reference Book,* pages 313 and 314.
This is a game for two players. Game materials include 1 deck each of *Fraction Of* Fraction Cards and Set Cards, the *Fraction Of* Gameboard, and a record sheet. This game provides practice with multiplication of fractions and whole numbers.

Fraction/Percent Concentration See *Student Reference Book,* page 315.
This game helps students memorize some of the easy fraction/percent equivalencies. Two or three players use 1 set of *Fraction/Percent Concentration* tiles and a calculator to play.

Fraction Top–It See *Student Reference Book,* page 316.
This game is for 2–4 players. Game materials include 1 deck of 32 Fraction Cards. This game provides practice with comparing fractions.

As You Help Your Child with Homework

As your child brings assignments home, you might want to go over the instructions together, clarifying them as necessary. The answers listed below will guide you through this unit's Study Links.

Study Link 5·1

1. 9   2. 14   3. $\frac{16}{20}$, or $\frac{4}{5}$

4. $\frac{45}{50}$, or $\frac{9}{10}$ 5. 70   6. 16

7. 9   8. a. $9   b. $20

c. Jen paid $\frac{2}{5}$ of the bill: 8 ÷ 2 = 4. So that means each fifth of the total was $4. Then $\frac{3}{5}$ must be $12. And $12 + $8 = $20.

9. 14   10. 140   11. 14   12. 140

Study Link 5·2

1. $2\frac{1}{2}$; $\frac{5}{2}$ 2. $2\frac{4}{6}$, or $2\frac{2}{3}$; $\frac{16}{6}$, or $\frac{8}{3}$

3. $1\frac{2}{3}$; $\frac{5}{3}$ 4. $2\frac{1}{6}$; $\frac{13}{6}$ 5. $2\frac{5}{6}$; $\frac{17}{6}$

7. 262   8. 32 R4   9. 123   10. 72 R3

Study Link 5·3

1. 4   2. 12   3. 1; 4

4. $\frac{4}{4}$ = 1   5. $\frac{6}{8}$ = $\frac{3}{4}$ 6. $\frac{5}{4}$ = $1\frac{1}{4}$

7. $\frac{9}{8}$, or $1\frac{1}{8}$ cups   9. 297

10. 148 R3   11. 74 R3   12. 37 R3

Study Link 5·4

1. =   2. ≠   3. ≠   4. =   5. =

6. =   7. =   8. =   9. 6   10. 21

11. 4   12. 40   13. 12   14. 80   15. 27

16. 56   17. 150   18. 70   19. $7.04

20. $20.03   21. 17 R10   22. 80 R4

Study Link 5·5

2. 0.4; 1.9; 20.7; 24.0; 60.9; 160.6; 181.3; 297.9; 316.0

Study Link 5·6

1. $7\frac{79}{100}$; $7\frac{78}{100}$, or $7\frac{39}{50}$; $6\frac{21}{100}$; $4\frac{7}{10}$; $3\frac{6}{10}$, or $3\frac{3}{5}$

2. a. $\frac{15}{45}$, or $\frac{1}{3}$ b. $\frac{9}{45}$, or $\frac{1}{5}$ c. $\frac{3}{45}$, or $\frac{1}{15}$

3. $0.\overline{3}$; 0.2; $0.0\overline{6}$ 4. 714 R6

5. 8 R4   6. 67 R5

Study Link 5·7

Sample answers given for Problem 1–5.

1. 0.25; 0.5; 0.75   2. 2.25; 2.5; 2.75

3. 0.65; 0.7; 0.775   4. 0.325; 0.35; 0.375

5. 0.051; 0.055; 0.059   6. 0.53

7. 0.2   8. 0.77   9. $0.\overline{8}$ 10. 0.051

11. 0.043; 0.05; 0.1; 0.12; 0.2; 0.6; 0.78

12. $7.06   13. 6 R17   14. 81   15. 694 R3

Study Link 5·8

1. $\frac{3}{4}$ = 0.75 = 75%; $\frac{14}{16}$ = 0.875 = 88%;

$\frac{15}{25}$ = 0.6 = 60%; $\frac{17}{20}$ = 0.85 = 85%;

$\frac{3}{8}$ = 0.375 = 38%

3. $\frac{3}{8}$; $\frac{15}{25}$; $\frac{3}{4}$; $\frac{17}{20}$; $\frac{14}{16}$ 4. $130   5. 10 questions

6. 97 R5   7. 48 R15   8. 32 R15   9. 24 R15

Study Link 5·9

2. Bar graph

3. Line graph; Temperature went up and down.

Study Link 5·10

1. a. 50%   b. 15%   c. 35%

3. 25% of the students in my class have skateboards. 25% have in-line skates. 50% have bicycles.

4. 633   5. 1.1636   6. 10 R1   7. 100 R4

Study Link 5·11

Check your child's circle graph.

2. 17   3. 23   4. 9   5. 7

Study Link 5·12

1. Mona ate 1 more cookie than Tomas. $\frac{3}{8}$ of 24 is 9; but $\frac{2}{5}$ of 25 is 10.

2. 12 students were sick. If $\frac{2}{3}$ is 24, that means $\frac{1}{3}$ is 12 students. So that means the rest of the class, or $\frac{1}{3}$ of the class, or 12 students, is sick.

4. 3   5. 24   6. 22   7. 24

STUDY LINK
5·1

Parts-and-Whole Fraction Practice

For the following problems, use counters or draw pictures to help you.

1. If 15 counters are the whole set, how many are $\frac{3}{5}$ of the set?

   _____ counters

2. If 18 counters are the whole set, how many are $\frac{7}{9}$ of the set? _____ counters

3. If 20 counters are the whole set, what fraction of the set is 16 counters? _____

4. If 50 counters are the whole set, what fraction of the set is 45 counters? _____

5. If 35 counters are half of a set, what is the whole set? _____ counters

6. If 12 counters are $\frac{3}{4}$ of a set, what is the whole set? _____ counters

7. Gerald and Michelle went on a 24-mile bike ride.
   By lunchtime, they had ridden $\frac{5}{8}$ of the total distance.

   How many miles did they have left to ride after lunch? _____ miles

8. Jen and Heather went to lunch. When the bill came, Jen discovered that she had only $8. Luckily, Heather had enough money to pay the other part, or $\frac{3}{5}$, of the bill.

   a. How much did Heather pay? _____

   b. How much was the total bill? _____

   c. Explain how you figured out Heather's portion of the bill.

   _____

   _____

   _____

Practice

9. 3)42 _____

10. 3)420 _____

11. 30)420 _____

12. 30)4,200 _____

STUDY LINK
5·2 | **Fraction and Mixed-Number Practice**

For the problems below, the hexagon is worth 1.
Write the mixed-number name and the fraction name
shown by each diagram.

Unit
hexagon

SRB
62 63

1. 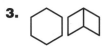 Mixed
number _____

Fraction _____

2. Mixed
number _____

Fraction _____

3.   Mixed
number _____

Fraction _____

4. Mixed
number _____

Fraction _____

5. Mixed
number _____

Fraction _____

6. Make up a mixed-number problem of your
own in the space below.

Practice

7. 7)‾1,834 _____

8. 6)‾196 → _____

9. 8)‾984 _____

10. 9)‾651 → _____

95

Fraction-Stick Problems

Shade the fraction sticks to help you find equivalent fractions.

1. $\frac{1}{2} = \frac{\square}{8}$

2. $\frac{3}{4} = \frac{\square}{16}$

3. $\frac{\square}{4} = \frac{2}{8} = \frac{\square}{16}$

Shade the fraction sticks to help you solve the addition problems.

4. $\frac{1}{4} + \frac{3}{4} =$ _____

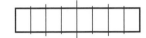

5. $\frac{1}{2} + \frac{2}{8} =$ _____

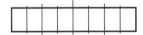 

6. $\frac{1}{2} + \frac{3}{4} =$ _____

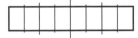 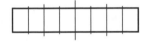

Shade the fraction sticks to help you solve the fraction number stories.

7. Joe was baking a cake. He added $\frac{3}{4}$ cup of white sugar and $\frac{3}{8}$ cup of brown sugar. How much sugar did he use in all?

_____
(unit)

8. On the back of this page, write a number story using fractions. Then write a number model to show how you solved it.

Practice

9. $3\overline{)891}$ _____

10. $6\overline{)891}$ $\rightarrow$ _____

11. $12\overline{)891}$ $\rightarrow$ _____

12. $24\overline{)891}$ $\rightarrow$ _____

97

STUDY LINK 5·4 | Equivalent Fractions

SRB
59

If the fractions are equivalent, write = in the answer blank.

If the fractions are not equivalent, write ≠ (not equal to) in the answer blank.

1. $\dfrac{3}{4}$ _____ $\dfrac{9}{12}$

2. $\dfrac{3}{10}$ _____ $\dfrac{1}{5}$

3. $\dfrac{7}{14}$ _____ $\dfrac{8}{15}$

4. $\dfrac{10}{12}$ _____ $\dfrac{5}{6}$

5. $\dfrac{16}{100}$ _____ $\dfrac{8}{50}$

6. $\dfrac{36}{72}$ _____ $\dfrac{1}{2}$

7. $\dfrac{7}{12}$ _____ $\dfrac{21}{36}$

8. $\dfrac{8}{3}$ _____ $\dfrac{16}{6}$

Fill in the boxes to complete and match the equivalent fractions.

Example: $\dfrac{\boxed{2}}{15} = \dfrac{6}{45}$

9. $\dfrac{3}{5} = \dfrac{\boxed{}}{10}$

10. $\dfrac{2}{3} = \dfrac{14}{\boxed{}}$

11. $\dfrac{44}{55} = \dfrac{\boxed{}}{5}$

12. $\dfrac{12}{\boxed{}} = \dfrac{3}{10}$

13. $\dfrac{35}{60} = \dfrac{7}{\boxed{}}$

14. $\dfrac{9}{16} = \dfrac{45}{\boxed{}}$

15. $\dfrac{9}{36} = \dfrac{\boxed{}}{108}$

16. $\dfrac{7}{\boxed{}} = \dfrac{1}{8}$

17. $\dfrac{30}{135} = \dfrac{\boxed{}}{27}$

18. $\dfrac{10}{16} = \dfrac{\boxed{}}{112}$

Practice

19. $7\overline{)\$49.28}$ _____

20. $15\overline{)\$300.45}$ _____

21. $21\overline{)367} \rightarrow$ _____

22. $8\overline{)644} \rightarrow$ _____

99

STUDY LINK 5·5 | Decimal Numbers

1. Mark each number on the number line. The first one is done for you.

   30.13   30.72   31.05   29.94   30.38

   30.13

   ```
   ◄──┼─┼─┼─┼─●─┼─┼─┼─┼─┼─┼─┼─┼─┼─┼─┼─┼─┼─┼─┼─┼─┼─┼──►
      29.9 30.0 30.1 30.2 30.3 30.4 30.5 30.6 30.7 30.8 30.9 31.0 31.1
   ```

2. Round the area of each country to the nearest tenth of a square kilometer.

	Ten Smallest Countries	Area in Square Kilometers	Area Rounded to the Nearest Tenth of a Square Kilometer
1	Vatican City	0.44 km²	km²
2	Monaco	1.89 km²	km²
3	Nauru	20.72 km²	km²
4	Tuvalu	23.96 km²	km²
5	San Marino	60.87 km²	km²
6	Liechtenstein	160.58 km²	km²
7	Marshall Islands	181.30 km²	km²
8	St. Kitts and Nevis	296.37 km²	km²
9	Maldives	297.85 km²	km²
10	Malta	315.98 km²	km²

Source: *The Top 10 of Everything 2005*

Practice

Solve and write the fact family number sentences.

3. $32\overline{)768}$

   _____ ÷ _____ = _____   _____ ÷ _____ = _____

   _____ * _____ = _____   _____ * _____ = _____

101

**STUDY LINK
5·6**

Decimals, Fractions, and Mixed Numbers

SRB
63
83 89

1. Convert each decimal measurement to a mixed number.

Longest Road and Rail Tunnels in the U.S.	Decimal Length	Mixed-Number Length
Cascade Tunnel (Washington)	7.79 miles	_____ miles
Flathead Tunnel (Montana)	7.78 miles	_____ miles
Moffat Tunnel (Colorado)	6.21 miles	_____ miles
Hoosac Tunnel (Massachusetts)	4.7 miles	_____ miles
BART Transbay Tubes (San Francisco, CA)	3.6 miles	_____ miles

Source: *The Top 10 of Everything 2005*

2. The longest one-word name of any place in America is Chargoggagoggmanchauggagoggchaubunagungamaugg.

   This name for a lake near Webster, Massachusetts, is 45 letters long. It is a Native American name that means "You fish on your side, I'll fish on mine, and no one fishes in the middle." Use this word to answer the problems below.

   a. What fraction of the word is made up of the letter *g*? _____ = _____

   b. What fraction of the word is made up of the letter *a*? _____ = _____

   c. What fraction of the word is made up of the letter *c*? _____ = _____

3. In the space above, write the decimal equivalents for the fractions in Problem 2.

Practice

4. $10\overline{)7,146}$ → _____

5. $10\overline{)84}$ → _____

6. $10\overline{)675}$ → _____

STUDY LINK 5·7 | Decimal Comparisons

Write three numbers between each pair of numbers.

1. 0 and 1 _____ , _____ , _____

2. 2 and 3 _____ , _____ , _____

3. 0.6 and 0.8 _____ , _____ , _____

4. 0.3 and 0.4 _____ , _____ , _____

5. 0.06 and 0.05 _____ , _____ , _____

Circle the correct answer to each question.

6. Which is closer to 0.6?   0.5   or   0.53

7. Which is closer to 0.3?   0.02   or   0.2

8. Which is closer to 0.8?   0.77   or   0.85

9. Which is closer to 0.75?   0.6   or $0.\overline{8}$

10. Which is closer to 0.04?   0.3   or   0.051

11. Arrange the decimals below in order from least to greatest.

0.12   0.05   0.2   0.78   0.6   0.043   0.1

_____ _____ _____ _____ _____ _____ _____

Practice

12. $9\overline{)\$63.54}$ _____

13. $45\overline{)287} \rightarrow$ _____

14. $7\overline{)567}$ _____

15. $7\overline{)4,861} \rightarrow$ _____

105

STUDY LINK 5·8   Percent Problems

1. Convert the following fractions to decimals and percents. Round to the nearest whole percent.

Fraction	Decimal	Percent
$\frac{3}{4}$		
$\frac{14}{16}$		
$\frac{15}{25}$		
$\frac{17}{20}$		
$\frac{3}{8}$		

2. On the back of this page, explain how you could find the percent equivalent to $\frac{17}{20}$ without using a calculator.

3. Write the five fractions from Problem 1 in order from least to greatest.

_____ _____ _____ _____ _____

4. Katie spent 50% of her money on shoes for soccer. The shoes cost $65. How much money did Katie start with? _____

5. Tom got 70% of the questions correct on a music test. If he got 7 questions correct, how many questions were on the test? _____

Practice

6. $10\overline{)975}$ → _____

7. $20\overline{)975}$ → _____

8. $30\overline{)975}$ → _____

9. $40\overline{)975}$ → _____

Graphs

Brenda's class made a list of their favorite colors. Here are the results.

Blue 8   Red 7   Yellow 3   Green 2   Other 4

1. Circle each graph that correctly represents the data above. (There may be more than one.)

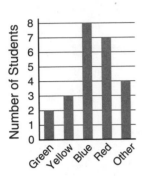
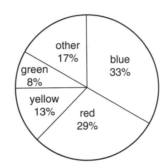
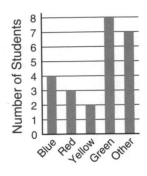

Marsha kept track of low temperatures. Here are the results for the end of May:

May 17	50°F	May 18	63°F	May 19	58°F	May 20	60°F
May 21	65°F	May 22	57°F	May 23	58°F	May 24	65°F
May 25	68°F	May 26	70°F	May 27	66°F	May 28	65°F
May 29	64°F	May 30	68°F	May 31	74°F		

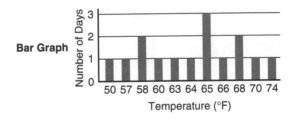

2. Which graph do you think is more helpful for answering the question, "On how many days was the low temperature 65°F?" _____

3. Which graph do you think is more helpful for showing trends in the temperature for the last two weeks of May? _____

4. On the back of this page, explain your choices for Problems 2 and 3.

STUDY LINK 5·10 | Circle Graphs and Collecting Data

SRB
125 126

1. Estimate the percent of the circle for each piece of the graph at the right.

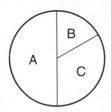

a. A is about _____ of the circle.

b. B is about _____ of the circle.

c. C is about _____ of the circle.

2. Draw a line connecting each data set with the most likely circle graph.

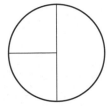 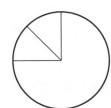 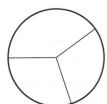 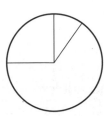

30% of Michel's class walks to school.	25% of Jeannene's toy cars are blue.	$\frac{1}{8}$ of Angelo's pants are jeans.
30% of Michel's class rides the bus.	10% of Jeannene's toy cars are striped.	$\frac{1}{8}$ of Angelo's pants are black dress pants.
40% of Michel's class rides in a car or van.	65% of Jeannene's toy cars are red.	$\frac{3}{4}$ of Angelo's pants are blue dress pants.

3. Circle the graph above that you did not use. Write a set of data to match that circle graph.

_____

_____

Practice

4. $6\overline{)3,798}$ _____

5. $7\overline{)8.145}$ _____

6. $2\overline{)21} \rightarrow$ _____

7. $8\overline{)804} \rightarrow$ _____

111

STUDY LINK 5·10 | Circle Graphs and Collecting Data *cont.*

SRB
125 126

The Number of States We've Been In

8. Talk with an adult at home and think of all the states you have visited. (Be sure to include the state you're living in.) Look at the map below to help you remember.

Use a pencil or crayon to mark each state you have visited.

Don't count any state that you have flown over in an airplane unless the plane landed, and you left the airport.

9. Count the number of states you have marked.

I have been in _____ states in my lifetime.

10. Now ask the adult to mark the map to show the states he or she has been in, using a different color or mark from yours.

Keep a tally as states are marked.

The adult I interviewed has visited _____ states.

Note: Alaska and Hawaii are not shown to scale.

Student and adult: This data is important for our next mathematics class.
Please bring this completed Study Link back to school tomorrow.

STUDY LINK 5·11 | What's in a Landfill?

People who study landfills have estimated the percent of landfill space (volume) taken up by paper, food, plastic, and so on.

Space in landfills taken up by:

Paper 50%

Food and yard waste 13%

Plastic 10%

Metal 6%

Glass 1%

Other waste 20%

> *Think of it this way:*
> For every 100 boxes of garbage hauled to the dump, expect that about 50 boxes could be filled with paper, 6 with metal, 1 with glass, and so on.

1. Cut out the Percent Circle. Use it to make a circle graph for the data in the table. (Remember to label the graph and give it a title.)

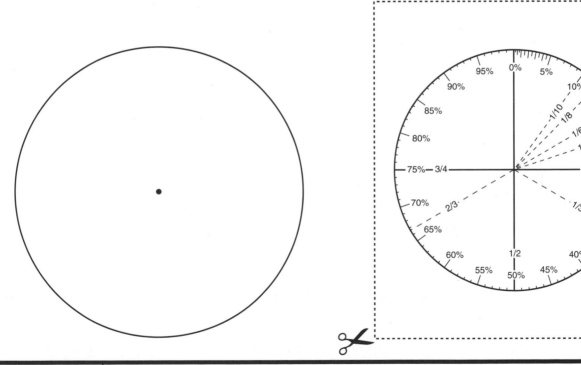

Practice

2. 23$\overline{)391}$ _____

3. 17$\overline{)391}$ _____

4. 43$\overline{)387}$ _____

5. 37$\overline{)259}$ _____

STUDY LINK 5·12 | Finding "Fractions of"

SRB
74

Solve.

1. Tomas ate $\frac{3}{8}$ of a bag of 24 cookies.
   Mona ate $\frac{2}{5}$ of a bag of 25 cookies.
   Who ate more cookies?
   Explain your answer.

   _____

   _____

   _____

   _____

   _____

   _____

2. On Thursday, 24 fifth-grade students
   came to school. That was only $\frac{2}{3}$ of the
   total class. The rest were home sick. How
   many students were sick?
   Explain your answer.

   _____

   _____

   _____

   _____

   _____

   _____

3. Mario was on a 21-mile hiking trail. He walked $\frac{3}{7}$ of the trail before stopping for
   lunch. How far did he walk before lunch? Explain your answer.

   _____

   _____

   _____

   _____

   _____

Practice

4. $52\overline{)156}$ _____

5. $24\overline{)576}$ _____

6. $13\overline{)286}$ _____

7. $22\overline{)528}$ _____

STUDY LINK 5·13   Unit 6: Family Letter

Using Data; Addition and Subtraction of Fractions

The authors of *Everyday Mathematics* believe that students should work substantially with data. Unit 6 is designed to present and teach relevant data skills and concepts, allowing your child ample opportunities to practice organizing and analyzing the data that he or she collects.

The data that your child collects at first will usually be an unorganized set of numbers. After organizing the data using a variety of methods, he or she will study the **landmarks** of the data. The following terms are called landmarks because they show important features of the data.

♦ The **maximum** is the largest data value observed.

♦ The **minimum** is the smallest data value observed.

♦ The **range** is the difference between the maximum and the minimum.

♦ The **mode** is the most popular data value—the value observed most often.

♦ The **median** is the middle data value observed.

♦ The **mean,** commonly known as the average, is a central value for a set of data.

At the end of the unit, students will demonstrate their skills by conducting a survey of their peers, gathering and organizing the data, analyzing their results, and writing a summary report.

Your child will continue the American Tour by studying Native American measurements for length and distance, based on parts of the body. Students will convert these body measures to personal measures by measuring their fingers, hands, and arms in both metric and U.S. customary units. In addition, your child will learn how to read a variety of contour-type maps, such as climate, precipitation, and growing-seasons maps.

Finally, students will explore addition and subtraction of fractions by using paper slide rules, a clock face, and fraction sticks. They will learn to find common denominators and apply this skill to add and subtract fractions with unlike denominators.

Please keep this Family Letter for reference as your child works through Unit 6.

Vocabulary

Important terms in Unit 6:

angle of separation In *Everyday Mathematics,* the angle measure between spread fingers. The figure shows the angle of separation between a person's thumb and first finger.

Angle of separation

common denominator Any number except zero that is a multiple of the denominators of two or more fractions. For example, the fractions $\frac{1}{2}$ and $\frac{2}{3}$ have common denominators 6, 12, 18, and so on.

contour line A curve on a map through places where a certain measurement (such as temperature or elevation) is the same. Often, contour lines separate regions that have been colored differently to show a range of conditions.

cubit An ancient unit of length, measured from the point of the elbow to the end of the middle finger. A cubit is about 18 inches.

decennial Occurring every 10 years.

fair game A game in which each player has the same chance of winning. If any player has an advantage or disadvantage, then the game is not fair.

fathom A unit used by people who work with boats and ships to measure depths underwater and lengths of cables. A fathom is now defined as 6 feet.

great span The distance from the tip of the thumb to the tip of the little finger (pinkie), when the hand is stretched as far as possible.

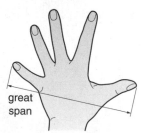

great span

landmark A notable feature of a data set. Landmarks include the *median, mode, maximum, minimum,* and *range.*

line plot A sketch of data in which check marks, Xs, or other marks above a labeled line show the frequency of each value.

map legend (map key) A diagram that explains the symbols, markings, and colors on a map.

mode The value or values that occur most often in a set of data.

normal span The distance from the tip of the thumb to the tip of the first (index) finger of an outstretched hand. Also called *span.*

population In data collection, the group of people or objects that is the focus of the study.

range The difference between the *maximum* and *minimum* in a set of data.

sample A part of a population chosen to represent the whole population.

simplest form A fraction less than 1 is in simplest form if there is no number other than 1 that divides its numerator and denominator evenly. A mixed number is in simplest form if its fractional part is in simplest form.

stem-and-leaf plot A display of data in which digits with larger place values are "stems" and digits with smaller place values are "leaves."
Data list: 24, 24, 25, 26, 27, 27, 28, 31, 31, 32, 32, 36, 36, 36, 41, 41, 43, 45, 48, 50, 52

Stem-and-leaf plot

Stems (10s)	Leaves (1s)
2	4 4 5 6 7 7 8
3	1 1 2 2 6 6 6
4	1 1 3 5 8
5	0 2

survey A study that collects data.

Do-Anytime Activities

To work with your child on the concepts taught in this unit and in previous units, try these interesting and rewarding activities.

1. Have your child design and conduct an informal survey. Help him or her collect and organize the data, and then describe the data using data landmarks. Challenge your child to create different ways to present the data.

2. Encourage your child to develop his or her own set of personal measures for both metric and U.S. customary units.

Building Skills through Games

In this unit, your child will work on his or her understanding of angles and the addition and subtraction of fractions by playing the following games. For detailed instructions, see the *Student Reference Book.*

Divisibility Dash See *Student Reference Book,* page 302. This is a game for two or three players. Game materials include 4 each of the number cards 0–9 as well as 2 each of the number cards 2, 3, 5, 6, 9, and 10. This game provides practice in recognizing multiples and using divisibility rules in a context that also develops speed.

Frac-Tac-Toe See *Student Reference Book,* pages 309–311. This is a game for two players. Game materials include 4 each of the number cards 0–10, pennies or counters of two colors, a calculator, and a gameboard. The gameboard is a 5-by-5 number grid that resembles a bingo card. Several versions of the gameboard are shown in the *Student Reference Book. Frac-Tac-Toe* helps students practice converting fractions to decimals and percents. In Unit 6, students practice fraction/decimal conversions.

Fraction Capture See *Math Journal,* page 198. This is a game for two players and requires 2 six-sided dice and a gameboard. Partners roll dice to form fractions and then attempt to capture squares on a *Fraction Capture* gameboard. This game provides practice in finding equivalent fractions and in adding fractions.

As You Help Your Child with Homework

As your child brings assignments home, you might want to go over the instructions together, clarifying them as necessary. The answers listed below will guide you through this unit's Study Links.

Study Link 6·1

3. a. 59   **b.** 24   **c.** 33

    d. 36   **e.** 39.5

5. 18.43   **6.** 16

Study Link 6·2

2. a. cm; ft   **b.** ounces; gal; liters

    c. m; miles   **d.** cm; ft; mm

    e. kg; lb; grams

3. 2,686   **6.** 141.63

Study Link 6·3

1. 73; maximum   **2.** 19   **3.** 53

4. Sample answer: Cross off the highest and lowest values—31 and 73. Continue by crossing off the highest and lowest values remaining, so that only one number, 53, remains.

5. 3,286   **8.** 65,250

Study Link 6·4

1. Tapes and CDs   **2.** Books and magazines

3. Movie tickets   **4.** 5,593

5. 16,539   **6.** 582 R3   **7.** 75,896

Study Link 6·5

Sample answers given for Problems 1–3.

1. 5, 7, 7, 8, 8, 9, 10, 13, 14, 15, 15, 15, 20

2.

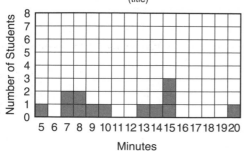

Minutes Needed to Get Ready for Bed
(title)

Number of Students (vertical axis)

Minutes (horizontal axis)

3. The number of minutes it takes to get ready for bed

5. 443   **7.** 1,839

Study Link 6·6

1. Sample answer: Ages of the oldest people we know
Title: The Oldest People Our Class Knows
Unit: Years

4. a. 32   **b.** 99   **c.** 66   **d.** 78.5

5. 12,495   **7.** 8,484

Study Link 6·7

1. California; Arizona   **2.** Montana; Washington

4. 2,086   **6.** 81

Study Link 6·8

1. $\frac{10}{14}$, or $\frac{5}{7}$ **3.** $\frac{6}{15}$, or $\frac{2}{5}$

5. 9,384   **7.** 2,952

Study Link 6·9

1. $\frac{22}{15}$, or $1\frac{7}{15}$ **2.** $\frac{1}{18}$

3. $\frac{9}{4}$, or $2\frac{1}{4}$ **4.** 4; $7\frac{3}{4}$

5. $5\frac{5}{6}$

Study Link 6·10

1. $\frac{18}{22} - \frac{11}{22} = \frac{7}{22}$ **2.** $\frac{20}{36} - \frac{9}{36} = \frac{11}{36}$

3. $\frac{21}{30} + \frac{8}{30} = \frac{29}{30}$ **4.** $\frac{21}{30} - \frac{8}{30} = \frac{13}{30}$

5. $\frac{19}{18}$, or $1\frac{1}{18}$ **6.** $\frac{59}{42}$, or $1\frac{17}{42}$

7. $\frac{1}{6}$ **8.** $\frac{3}{4}$

9. $\frac{2}{12}$, or $\frac{1}{6}$ **10.** $\frac{1}{2}$

11. $\frac{1}{3}$ **12.** $\frac{23}{12}$, or $1\frac{11}{12}$

13. $\frac{23}{12}$, or $1\frac{11}{12}$ **14.** $\frac{19}{12}$, or $1\frac{7}{12}$

The Standing Long Jump

Ms. Perez's physical education class participated in the standing long jump. Following are the results rounded to the nearest inch.

24	35	33	48	33	48	27	35	27	55	43	24
55	33	52	33	29	59	26	59	48	37	42	42

1. Organize these data on the line plot below.

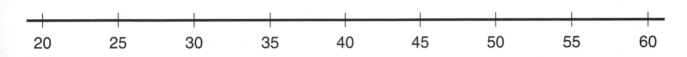

20   25   30   35   40   45   50   55   60

2. Make a bar graph for these data.

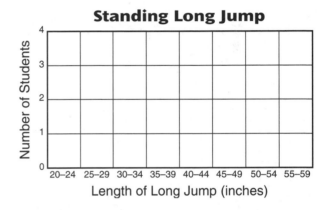

Standing Long Jump

Number of Students — Length of Long Jump (inches)
20–24 25–29 30–34 35–39 40–44 45–49 50–54 55–59

3. Find the following landmarks for the standing long jump data:

a. Maximum: _____ in.   **b.** Minimum: _____ in.

c. Mode: _____ in.   **d.** Median: _____ in.

e. Mean (average): _____ in. (Use a calculator. Add the distances and divide the sum by the number of jumps. Round to the nearest tenth.)

Practice

4. 48 * 29 = _____

5. 98.25
   − 79.82

6. 24)‾384‾

7. 767.5 + 30.82 = _____

121

Standard and Nonstandard Units

1. Use your body measures to find three objects that are about the size of each measurement below.

a. 1 cubit

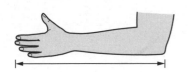

b. 1 great span

great span

c. 1 finger width

_____ _____ _____

_____ _____ _____

_____ _____ _____

2. For each problem below, mark the unit or units you *could* use to measure the object.

a. Height of your ceiling   ○ cm   ○ ft   ○ lb   ○ miles

b. Amount of milk in a pitcher   ○ cm   ○ ounces   ○ gal   ○ liters

c. Depth of the ocean   ○ m   ○ ounces   ○ gal   ○ miles

d. Length of a bee   ○ cm   ○ ft   ○ mm   ○ liters

e. Weight of a nickel   ○ in.   ○ kg   ○ lb   ○ grams

Practice

3. $34 * 79 =$ _____

4.   8,201
    $-2,190$

5. $6)\overline{4,152}$

6. $59.46 + 82.17 =$ _____

STUDY LINK 6·3   Reading a Stem-and-Leaf Plot

Use the information below to answer the questions.

Jamal was growing sunflowers. After eight weeks, he measured the height of
his sunflowers in inches. He recorded the heights in the stem-and-leaf plot below.

SRB
118 119

1. How tall is the tallest sunflower? _____ in.

Which landmark is the height of the tallest
flower? Circle its name.

minimum   mode

maximum   mean

2. How many sunflowers did Jamal measure? _____ sunflowers.

3. What is the mode for his measurements? _____ in.

4. Explain how to find the median for his measurements.

_____

_____

_____

_____

_____

Height of Sunflowers (inches)

Stems (10s)	Leaves (1s)
3	9 1
4	7 6 9 2 9
5	2 3 3 5 2 8 7 3
6	5 3 4
7	3

Practice

5. 62 * 53 = _____

6.   6,711
      − 4,140

7. 22)398 → _____

8. 725 * 90 = _____

STUDY LINK 6·4 | # How Much Do Students Spend?

A fifth-grade class collected data about class spending per month on various items. Below are some of the results.

◆ A median amount of $6 per month was spent for books and magazines.

◆ A median amount of $10 per month was spent for tapes and CDs.

◆ A median amount of $8 per month was spent for movie tickets.

The number-line plots below display the data. Match the plots with the items: books and magazines, tapes and CDs, and movie tickets.

1. _____

```
                                        X   X
                                        X   X
                                    X   X   X
                          X   X     X   X   X   X                   X
    ────────────────────────────────────────────────────────────────────
       1   2   3   4   5   6   7   8   9  10  11  12  13  14  15  16
```

2. _____

```
                                X
            X       X       X   X
            X       X   X   X   X   X           X
    ────────────────────────────────────────────────────────────────────
       1   2   3   4   5   6   7   8   9  10  11  12  13  14  15  16
```

3. _____

```
                            X           X
                            X           X
                    X   X   X   X   X   X
                    X   X   X   X   X   X   X
    ────────────────────────────────────────────────────────────────────
       1   2   3   4   5   6   7   8   9  10  11  12  13  14  15  16
```

| **Practice** | |

4. 119 ∗ 47 = _____

5.   9,402
    + 7,137

6. 9)‾5,241 → _____

7. 9,487 ∗ 8 = _____

127

STUDY LINK 6·5 — Constructing a Graph from Landmarks

1. Make up a list of data with the following landmarks:

   mode: 15   minimum: 5   median: 10   maximum: 20

   Use at least 10 numbers.

   _____

2. Draw and label a bar graph to represent your data.

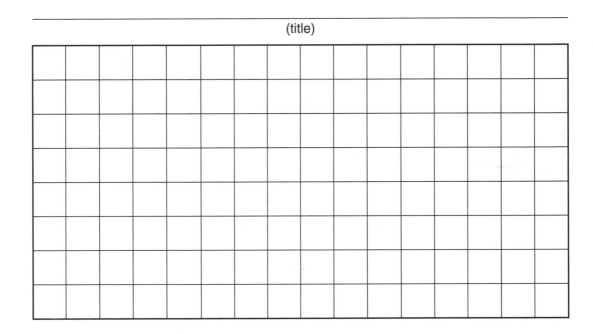

(title)

3. Describe a situation in which these data might actually occur.

   _____

   _____

Practice

4. 305 * 29 = _____

5. 524 − 81 = _____

6. 671 * 132 = _____

7. 7,356 ÷ 4 = _____

STUDY LINK 6·6 Data Analysis

1. Describe a situation in which the data in the line plot below might occur. Then give the plot a title and a unit.

_____

| _____ | _____ |
| (title) | (unit) |

```
                                    X
                              X   X                 X
          X         X       X  X  X  X      X       X
    X     X         X  X  X  X  X  X     X       X  X      X   X
    +---+---+---+---+---+---+---+---+---+---+---+---+---+---+---+---+---+---+
   77  78  79  80  81  82  83  84  85  86  87  88  89  90  91  92  93  94
```

2. Find the following landmarks for the data in the line plot.

   a. minimum: _____ b. maximum: _____ c. mode: _____ d. median: _____

3. Describe a situation in which the data in the stem-and-leaf plot shown below might occur. Then give the plot a title and a unit.

   _____

   _____

4. Find the following landmarks for the data in the stem-and-leaf plot.

   a. minimum: _____ b. maximum: _____

   c. mode: _____ d. median: _____

	_____
	(title)
	_____
	(unit)
Stems	**Leaves**
(10s)	**(1s)**
3	2
4	0
5	1 3 7
6	0 4 5 6 6 6 7 9
7	1 3 8 8 9
8	0 2 2 5 5 8 8 9
9	0 2 2 5 5 8 9 9

Practice

5. 245 * 51 = _____

6. 764 + 37 = _____

7. 2,121 * 4 = _____

8. 1,976 ÷ 38 = _____

131

Name   Date   Time

STUDY LINK
6·7 | **Contour Map**

Study the map below to answer the questions.

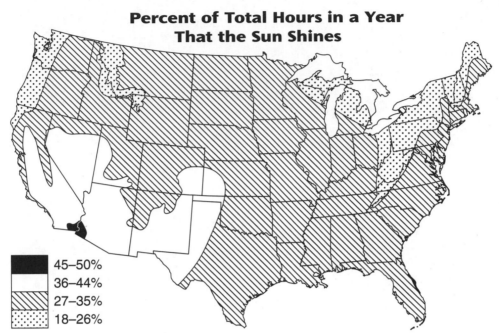

**Percent of Total Hours in a Year
That the Sun Shines**

45–50%
36–44%
27–35%
18–26%

1. States where at least part of the state has sunny days more than 45% of the time.

○ Washington   ○ California   ○ Arizona   ○ New York

2. States that border Canada where at least some part of the state has days that are NOT sunny at least 31% of the time.

○ California   ○ Montana   ○ Nebraska   ○ Washington

3. Make up your own question about the map. Answer your question.

_____

_____

| **Practice** |

4. 149 * 14 = _____   **5.** 134 * 29 = _____

6. 2,997 ÷ 37 = _____   **7.**   3,682
                      −1,590

133

STUDY LINK 6·8 — Calculating with Fraction Sticks

Solve. Use the fraction sticks to help you.

SRB
68–70

1. $\dfrac{3}{7} + \dfrac{4}{14} =$ _____

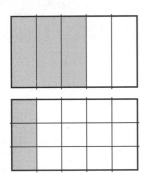

2. $1\dfrac{1}{2} + 2\dfrac{3}{4} =$ _____

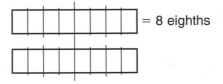

3. $\dfrac{3}{5} - \dfrac{3}{15} =$ _____

4. Write an open number sentence and solve. Shade in the fraction stick to help you.

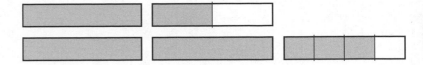 = 8 eighths

_____

Practice

Show your work.

5. $408 * 23 =$ _____

6. $0.85 + 0.3 =$ _____

7. $492 * 6 =$ _____

8. $45\overline{)2{,}297} \rightarrow$ _____

135

STUDY LINK 6·9

Adding and Subtracting Fractions

Multiplication Rule

To find a fraction equivalent to a given fraction, multiply the numerator and the denominator of the fraction by the same number.

$$\frac{a}{b} = \frac{a * n}{b * n}$$

Example 1: $\frac{4}{9} - \frac{1}{3} = ?$

$$\frac{1}{3} = \frac{2}{6} = \boxed{\frac{3}{9}} = \frac{4}{12} = \frac{5}{15} = \frac{6}{18} = \dots$$

9 is a common denominator.

$$\frac{4}{9} - \frac{1}{3} = \frac{4}{9} - \frac{3}{9} = \frac{1}{9}$$

Example 2: $\frac{5}{8} + \frac{2}{5} = ?$

$$\frac{5}{8} = \frac{10}{16} = \frac{15}{24} = \frac{20}{32} = \boxed{\frac{25}{40}} = \frac{30}{48} = \dots$$

$$\frac{2}{5} = \frac{4}{10} = \frac{6}{15} = \frac{8}{20} = \frac{10}{25} = \frac{12}{30} = \frac{14}{35} = \boxed{\frac{16}{40}} = \frac{18}{45} = \dots$$

Both fractions can be rewritten with the common denominator 40.

$$\frac{5}{8} + \frac{2}{5} = \frac{25}{40} + \frac{16}{40} = \frac{41}{40}, \text{ or } 1\frac{1}{40}$$

Find a common denominator. Then add or subtract.

1. $\frac{2}{3} + \frac{4}{5} =$ _____

2. $\frac{8}{9} - \frac{5}{6} =$ _____

3. $\frac{3}{4} + 1\frac{1}{2} =$ _____

4. Lisa was 4 feet $10\frac{1}{2}$ inches tall at the end of fifth grade. During the year, she had grown $2\frac{3}{4}$ inches. How tall was Lisa at the start of fifth grade?

_____ feet _____ in.

5. Bill was baking two different kinds of bread. One recipe called for $3\frac{1}{2}$ cups of flour. The other called for $2\frac{1}{3}$ cups of flour. How much flour did Bill need in all?

_____ cups

137

STUDY LINK 6·10 | **Fractions**

Find a common denominator. Then add or subtract.

1. $\frac{9}{11} - \frac{1}{2} =$ _____

2. $\frac{5}{9} - \frac{1}{4} =$ _____

3. $\frac{7}{10} + \frac{4}{15} =$ _____

4. $\frac{7}{10} - \frac{4}{15} =$ _____

5.
$$\begin{array}{r} \frac{3}{2} \\ - \frac{4}{9} \end{array}$$

6.
$$\begin{array}{r} \frac{5}{6} \\ + \frac{4}{7} \end{array}$$

Write the fraction represented by the shaded part of each fraction stick.

7. 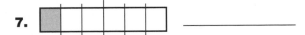 _____

8.  _____

9.  _____

10. 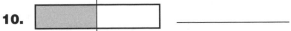 _____

11. _____

12. The sum of the five fractions in Problems 7–11 is _____ .

Use the information on Kwame's shopping list to fill in the blanks below.

13. He plans to buy _____ pounds of meat.

14. He plans to buy _____ pounds of cheese.

> **Kwame's Shopping List**
>
> $\frac{1}{2}$ pound ham
>
> $\frac{3}{4}$ pound roast beef
>
> $\frac{2}{3}$ pound turkey
>
> $\frac{2}{3}$ pound Swiss cheese
>
> $\frac{1}{4}$ pound Parmesan cheese
>
> $\frac{2}{3}$ pound cheddar cheese

STUDY LINK
6·11

Unit 7: Family Letter

Exponents and Negative Numbers

In Unit 7, your child will learn to write exponential and scientific notation for naming very large and very small numbers. These topics become increasingly important later on when your child begins algebra. If you have enjoyed playing math games in the past, you might want to play *Exponent Ball* during these lessons.

Your child will also review how parentheses make expressions unambiguous and will learn rules that determine the order for performing operations in a mathematical expression.

Finally, your child will learn to work with positive and negative numbers, using a variety of tools. For example, your child will use number lines, a slide rule, and red and black "counters" to model addition and subtraction problems.

The counter activities are especially helpful. Students use counters to represent an account balance. The red counters (−$1) represent a debit, and the black counters (+$1) represent a credit. If there are more red counters than black ones, the account is "in the red," that is, the balance is negative. On the other hand, if there are more black counters than red ones, the account is "in the black," that is, the balance is positive. By adding or subtracting red and black counters from an account, your child can model addition and subtraction of positive and negative numbers. To assist your child, you might want to explain how a checking or savings account works. Students will practice their new skills in the *Credits/Debits Game*.

Please keep this Family Letter for reference as your child works through Unit 7.

141

Vocabulary

Important terms in Unit 7:

account balance An amount of money that you have or that you owe.

exponential notation A way to show repeated multiplication by the same factor. For example, 2^3 is exponential notation for $2 * 2 * 2$.

expression A mathematical phrase made up of numbers, variables, operation symbols, and/or grouping symbols. An expression does not contain symbols such as $=$, $>$, and $<$.

in the black Having a positive balance; having more money than is owed.

in the red Having a negative balance; owing more money than is available.

negative number A number less than zero.

nested parentheses Parentheses within parentheses in an *expression*. Expressions are evaluated from within the innermost parentheses outward following the *order of operations*.

Example:

$((6 * 4) - 2) / 2$
$(24 - 2) / 2$
$22 / 2 = 11$

number-and-word notation A way of writing a large number using a combination of numbers and words. For example, 27 *billion* is number-and-word notation for 27,000,000,000.

opposite of a number A number that is the same distance from 0 on the number line as a given number but on the opposite side of 0. For example, the opposite of $+3$ is -3; the opposite of -5 is $+5$.

order of operations Rules that tell the order in which operations in an *expression* should be carried out. The order of operations is:

1. Do operations inside grouping symbols first. (Use rules 2–4 inside the grouping symbols.)
2. Calculate all the expressions with exponents.
3. Multiply and divide in order from left to right.
4. Add and subtract in order from left to right.

parentheses ( ) Grouping symbols used to indicate which operations in an expression should be done first.

scientific notation A system for writing numbers in which a number is written as the product of a power of 10 and a number that is at least 1 and less than 10. Scientific notation allows you to write big and small numbers with only a few symbols. For example, $4 * 10^{12}$ is scientific notation for 4,000,000,000,000.

slide rule An *Everyday Mathematics* tool for adding and subtracting integers and fractions.

standard notation Our most common way of representing whole numbers, integers, and decimals. Standard notation is base-ten, place-value numeration. For example, standard notation for three hundred fifty-six is 356.

Do-Anytime Activities

To work with your child on the concepts taught in this unit and in previous units, try these interesting and rewarding activities:

1. Have your child pick out a stock from the stock-market pages of a newspaper. Encourage your child to watch the stock over a period of time and to report the change in stock prices daily, using positive and negative numbers.

2. Using the same stock in Activity 1, have your child write the high and low of that stock for each day. After your child has watched the stock over a period of time, have him or her find. . .

   ◆ the *maximum* value observed.   ◆ the *mode,* if there is one.

   ◆ the *minimum* value observed.   ◆ the *median* value observed.

   ◆ the *range* in values.

3. Review tessellations with your child. Encourage your child to name the regular tessellations and to draw and name the 8 semiregular tessellations. Challenge your child to create Escher-type translation tessellations. You might want to go to the library first and show your child examples of Escher's work.

4. Practice finding perimeters of objects and circumferences of circular objects around your home.

Building Skills through Games

In Unit 7, your child will practice operations and computation skills by playing the following games. For detailed instructions, see the *Student Reference Book.*

Credits/Debits Game See *Student Reference Book,* page 301. Two players use a complete deck of number cards, cash and debt cards, and a record sheet to tally a balance. This game helps students add and subtract signed numbers.

Exponent Ball See *Student Reference Book,* page 305. This game involves two players and requires a gameboard, 1 six-sided die, a penny or counter, and a calculator. This game develops skills dealing with forming and comparing exponential values.

Name That Number See *Student Reference Book,* page 325. This is a game for two or three players using the Everything Math Deck or a complete deck of number cards. Playing *Name That Number* helps students review operations with whole numbers.

Scientific-Notation Toss See *Student Reference Book,* page 329. Two players will need 2 six-sided dice to play this game. This game develops skill in converting numbers from scientific notation to standard notation.

As You Help Your Child with Homework

As your child brings assignments home, you might want to go over the instructions together, clarifying them as necessary. The answers listed below will guide you through this unit's Study Links.

Study Link 7·1

2. Should be $6^3 = 6 * 6 * 6$; 216

3. Should be $2^9 = 2 * 2 * 2 * 2 * 2 * 2 * 2 * 2 * 2$; 512

4. Should be $4^7 = 4 * 4 * 4 * 4 * 4 * 4 * 4$; 16,384

5. 14.7   6. 0.48   7. $\frac{15}{7}$, or $2\frac{1}{7}$

Study Link 7·2

1. billion   2. 10^3 3. trillion

4. 10^6 5. thousand; 10^3 6. million; 10^6

7. $2^4 * 3$ 8. $2^2 * 3 * 5$

9. $3,000 + 200 + 60 + 4$

Study Link 7·3

1. 600; 3   2. 6   3. 500 million

4. 260 million 5. 10 million   6. 125

Study Link 7·4

1. $2 = (3 * 2) - (4 / 1)$ 2. $3 = (4 + 3 - 1) / 2$

3. $4 = (3 - 1) + (4 / 2)$ 5. $1 = ((4 + 1) - 3) / 2$

6. $6 = (1 + (4 * 2)) - 3$

7. $(4^2 - ((3 * 3)) + 1((2 + 1)^4 \div 9) - 1$

8. $a = 1\frac{4}{12}$, or $1\frac{1}{3}$ 9. $p = 1\frac{1}{2}$

10. $d = 2\frac{2}{8}$, or $2\frac{1}{4}$ 11. $y = 0$

Study Link 7·5

1. 34   2. 25   3. 28   4. 30

5. 21   6. 28   7. false   8. true

9. true   10. true   11. false   12. true

13. false   14. true   15. $z = 9,204$

16. $r = 78,002$ 17. $s = 1.25$

Study Link 7·6

1. Sales were at their highest in 1930. Sales dropped by 60 million from 1940 to 1970.

3. Before TV sets were common, more people went to the movies.

Study Link 7·7

1. 2.6   2. 1.58   3. −5.5

4. −9.8   5. −1.2, −1, 3.8, $5\frac{1}{4}$, $5\frac{3}{8}$

7. F   8. F   9. T

10. T   11. $-1 < 1$; T 13. $f = 12.53$

15. $n = \frac{3}{4}$

Study Link 7·8

1. <   2. >   3. >   4. >

5. 2 debt   6. 5 cash   7. 9   9. −88

11. 3   15. $a = 30$ 17. $p = 5$

Study Link 7·9

1. −41   2. 43   3. 0   4. −8

5. 40   6. 20   7. −85   8. −0.5

9. 2   10. (−10) 12. $u = 65,664$

13. $e = 3$ 14. $w = 30.841$ 15. $m = 5.46$

Study Link 7·10

1. <   2. >   3. >   4. >

5. >   6. >   7. −5   8. −21

9. 4   10. −6   11. −11   12. −26

13. 16   14. −4   15. true

16. true   17. $(-2 + 3) * 4 = 4$

Study Link 7·11

1. $-5 - (-58) = 53$ 3. 10^4

7. 20,000   13. $7 * 10^9$ 19. $b = 0.46$

21. $a = 1,571$ 23. $137\frac{4}{7}$, or 137 R4

Exponents

In exponential notation, the **exponent** tells how many times the **base** is used as a factor. For example, $6^4 = 6 * 6 * 6 * 6 = 1{,}296$. The base is 6, and the exponent is 4. The product is written as 1,296 in standard notation.

1. Complete the table.

Exponential Notation	Base	Exponent	Repeated Factors	Standard Notation
9^3	9	3	$9 * 9 * 9$	729
	4	5		
			$7 * 7 * 7 * 7$	
			$10 * 10 * 10 * 10 * 10 * 10$	
				262,144

Describe the mistake. Then find the correct solution.

2. $6^3 = 6 + 3 = 9$

    Mistake: _____

    Correct solution: _____

3. $2^9 = 9 + 9 = 18$

    Mistake: _____

    Correct solution: _____

4. $4^7 = 4 * 7 = 28$

    Mistake: _____

    Correct solution: _____

Practice

5. $351.82 + n = 366.52$ 6. $100 - r = 99.52$ 7. $\frac{4}{7} + u = \frac{19}{7}$

    $n =$ _____ $r =$ _____ $u =$ _____

Guides for Powers of 10

There are prefixes that name powers of 10. You know some of them from the metric system. For example, *kilo-* in kilometer (1,000 meters). It's helpful to memorize the prefixes for every third power of 10 through one trillion.

Memorize the table below. Have a friend quiz you. Then cover the table, and try to complete the statements below.

Standard Notation	Number-and-Word Notation	Exponential Notation	Prefix
1,000	1 thousand	10^3	kilo-
1,000,000	1 million	10^6	mega-
1,000,000,000	1 billion	10^9	giga-
1,000,000,000,000	1 trillion	10^{12}	tera-

1. More than 10^9, or one _____, people live in China.

2. One thousand, or $10^{\boxed{}}$, feet is a little less than $\frac{1}{5}$ of a mile.

3. Astronomers estimate that there are more than 10^{12}, or one _____, stars in the universe.

4. More than one million, or $10^{\boxed{}}$, copies of *The New York Times* are sold every day.

5. A kiloton equals one _____, or $10^{\boxed{}}$, metric tons.

6. A megaton equals one _____, or $10^{\boxed{}}$, metric tons.

Practice

Find the prime factorization of each number, and write it using exponents.

7. 48 = _____   8. 60 = _____

Write each number in expanded notation.

9. 3,264 = _____

10. 675,511 = _____

STUDY LINK
7·3

Interpreting Scientific Notation

Scientific notation is a short way to represent large and small numbers. In scientific notation, a number is written as the product of two factors. One factor is a whole number or a decimal. The other factor is a power of 10.

Scientific notation: $4 * 10^4$

     Meaning: Multiply 10^4 (10,000) by 4.

                 $4 * 10^4 = 4 * 10,000 = 40,000$

Number-and-word notation: 40 thousand

Scientific notation: $6 * 10^6$

      Meaning: Multiply 10^6 (1,000,000) by 6.

                 $6 * 10^6 = 6 * 1,000,000 = 6,000,000$

Number-and-word notation: 6 million

Guides for Powers of 10	
10^3	one thousand
10^6	one million
10^9	one billion
10^{12}	one trillion

Complete the following statements.

1. The area of Alaska is about $6 * 10^5$, or _____ thousand, square miles.

     The area of the lower 48 states is about $3 * 10^6$, or _____ million, square miles.

2. There are about $6 * 10^9$, or _____ billion, people in the world.

3. It is estimated that about $5 * 10^8$, or _____, people speak English as their first or second language.

4. In Bengal, India, and Bangladesh there are about $2.6 * 10^8$, or _____, people who speak Bengali.

5. At least 1 person in each of $1 * 10^7$ households, or _____, watches the most popular TV shows.

Source: *The World Almanac and Book of Facts, 2000*

Practice

6. $5 * (3^2 + 4^2) =$ _____

7. $3 * (9 + 16) =$ _____

8. $2 * (9 + h) = 20$ _____

9. $g = (7^2 - 2^2)$ _____

STUDY LINK 7·4 Using Parentheses

Make each sentence true by inserting parentheses.

1. $2 = 3 * 2 - 4 / 1$ **2.** $3 = 4 + 3 - 1 / 2$ **3.** $4 = 3 - 1 + 4 / 2$

4. Write seven names for 8. Use only numbers less than 10, and use at least three different operations in each name. Use parentheses. Follow the directions in Problem 7 to fill in the last two rows.

8

Make each sentence true by inserting parentheses.

> **Reminder:** When you have a pair of parentheses inside another pair, the parentheses are called **nested parentheses**.
>
> **Example:** $8 = ((5 * 6) + 2) / 4$

5. $1 = 4 + 1 - 3 / 2$ **6.** $7 = 4 * 3 / 2 + 1$

7. Add two names to your name-collection box in Problem 4. Use nested parentheses.

Practice

Find the number that each variable represents.

8. $2\frac{5}{12} = (1\frac{1}{12} + a)$ _____   **9.** $(1\frac{1}{2} + p) * 2^2 = 12$ _____

10. $6\frac{5}{8} + d = 7\frac{15}{8}$ _____   **11.** $6.4 - y = 6\frac{2}{5}$ _____

STUDY LINK 7·5 Order of Operations

Rules for Order of Operations

1. Do operations inside **parentheses.**
2. Calculate all expressions with **exponents.**
3. **Multiply** and **divide** in order, from left to right.
4. **Add** and **Subtract** in order, from left to right.

SRB
223

Solve.

1. $4 + 5 * 6 =$ _____

2. $(2 + 3)^2 =$ _____

3. $12 * 2 + 8 \div 2 =$ _____

4. $115 - 10^2 + 3 * 5 =$ _____

5. $6 * (3 + 2^2) \div 2 =$ _____

6. $7 + 9 * 7 \div 3 =$ _____

Write true or false for each number sentence. Follow the rules for order of operations.

7. $3 + 4 * 5 = 35$ _____

8. $(3 + 4) * 5 = 35$ _____

9. $0 = 3 * 4 - 12$ _____

10. $0 = (3 * 4) - 12$ _____

11. $36 = 12 - 3 * 4$ _____

12. $36 = (12 - 3) * 4$ _____

13. $8 \div 2 + 6 = 1$ _____

14. $8 \div (2 + 6) = 1$ _____

Practice

Find the number that each variable represents.

15. $354 * 26 = z$ _____

16. $907 * 86 = r$ _____

17. $3.000 - 1.75 = s$ _____

18. $0.006 + 3.2 + 0.75 + 4 = h$ _____

STUDY LINK
7·6

Making Line Graphs

Bar graphs, circle graphs, and line graphs display information in a way that
makes it easy to show comparisons, but line graphs can also show trends.

1. Use the information in the line
   graph to write two true statements
   about movie ticket sales.

   _____

   _____

   _____

   _____

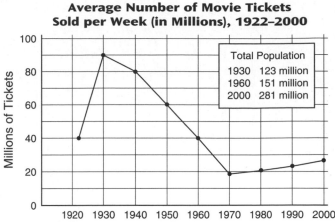

**Average Number of Movie Tickets
Sold per Week (in Millions), 1922–2000**

Total Population
1930   123 million
1960   151 million
2000   281 million

2. The table data lists the estimated percent of households with television sets
   from 1940 to 2000. Plot the data on the line graph below.

Estimated Percent of Households with Television Sets, 1940–2000						
Year	1940	1950	1960	1970	1980	1990	2000
Percentage	0%	12%	88%	96%	98%	98%	98%

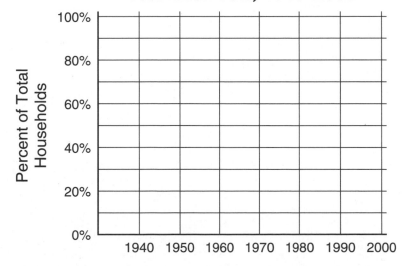

**Estimated Percent of Households with
Television Sets, 1940–2000**

3. Compare the information in the line graphs from Problems 1 and 2. What
   relationships do you see?

   _____

   _____

STUDY LINK
7·7

Greater Than or Less Than?

Name a number between each pair of numbers.

1. 2 and 3 _____

2. 1.5 and 2 _____

3. −5 and −6 _____

4. −9.5 and −10 _____

Order each set of numbers from *least* to *greatest*.

5. $5\frac{1}{4}$, 3.8, −1.2, −1, $5\frac{3}{8}$ _____

6. −6, $-4\frac{1}{2}$, −0.5, −7, 0 _____

True or false? Write T for true and F for false.

7. −6 > 5 _____

8. $5\frac{1}{2} < 5\frac{3}{6}$ _____

9. −2.5 > −3.5 _____

10. −4 is less than 0 _____

Write one true and one false number sentence. In each sentence,
use at least one negative number and one of the >, <, or = symbols.
Label each sentence T or F.

11. _____ _____

12. _____ _____

Practice

Find the number that each variable represents.

13. 92.47 + f = 105 _____

14. $32 + 15 + 25 + 8 + s = 10^2$ _____

15. $4\frac{3}{12} + n = 5$ _____

16. $4\frac{3}{12} - r = 3\frac{6}{12}$ _____

STUDY LINK 7·8 Positive and Negative Numbers

Write $<$ or $>$.

1. -7 _____ 6

2. 0.01 _____ -32

3. 8.5 _____ -10^3

4. $-\dfrac{3}{4}$ _____ -1.6

Find the account balance. $\boxed{+}$ = \$1 cash. $\boxed{-}$ = \$1 debt.

5. Balance = \$ _____

6. Balance = \$ _____

Solve these addition problems.

7. $-15 + 6 = $ _____

8. $17 + (-5) = $ _____

9. $-56 + (-32) = $ _____

10. $90 + (-20) = $ _____

11. $18 + (-15) = $ _____

12. $-987 + 987 = $ _____

13. Use the rule to complete the table.

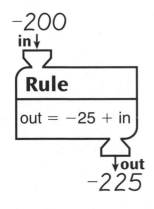

in	out
25	
50	
-25	
-100	
100	
0	

Practice

Find the number that each variable represents.

14. $3\dfrac{2}{3} = \dfrac{j}{3}$ _____

15. $7\dfrac{9}{3} = \dfrac{a}{3}$ _____

16. $\dfrac{19}{25} * \dfrac{y}{y} = \dfrac{57}{75}$ _____

17. $\dfrac{75}{100} \div \dfrac{p}{p} = \dfrac{15}{20}$ _____

STUDY LINK 7·9 | Addition and Subtraction Problems

Solve each problem. Be careful. Some problems involve addition, and some involve subtraction.

| Reminder: |
| To subtract a number, you can add the opposite of that number. |

1. $-25 + (-16) =$ _____

2. $0 - (-43) =$ _____

3. $-4 - (-4) =$ _____

4. $-4 - 4 =$ _____

5. $29 - (-11) =$ _____

6. $9 - (-11) =$ _____

7. $-100 + 15 =$ _____

8. $10 - 10.5 =$ _____

9. $4\frac{1}{2} + (-2\frac{1}{2}) =$ _____

10. $10 -$ _____ $= 20$

11. For each temperature change in the table, two number models are shown in the Temperature after Change column. Only one of the number models is correct. Cross out the incorrect number model. Then complete the correct number model.

Temperature before Change	Temperature Change	Temperature after Change	
40°	up 7°	$40 + 7 =$ _____	$40 + (-7) =$ _____
10°	down 8°	$10 - (-8) =$ _____	$10 - 8 =$ _____
−15° (15° below zero)	up 10°	$-15 + 10 =$ _____	$15 + 10 =$ _____
−20° (20° below zero)	down 10°	$-20 - 10 =$ _____	$20 - (-10) =$ _____

Practice

Find the number that each variable represents.

12. $684 * 96 = u$ _____

13. $69 \div e = 23$ _____

14. $32.486 - 1.645 = w$ _____

15. $9.45 - m = 3.99$ _____

Positive and Negative Number Review

Write >, < or =.

1. -8 _____ 5   2. -3 _____ -10 3. 10 _____ -20

4. 12 _____ -15 5. $-\dfrac{3}{4}$ _____ -1 6. 3^2 _____ 6

Add or subtract.

7. $-20 + 15 =$ _____   8. $-14 + (-7) =$ _____

9. $-8 + 12 =$ _____   10. $3 + (-9) =$ _____

11. $-4 - 7 =$ _____   12. $-10 - 16 =$ _____

13. $5 - (-11) =$ _____   14. $8 - 12 =$ _____

Some of the following number sentences are true because they follow the rules for the order of operations. Some of the sentences are false. Make a check mark next to the true number sentences. Insert parentheses in the false number sentences to make them true.

15. $3 + 7 * 5 = 38$ 16. $-5 + 20 \div 5 = -1$

17. $-2 + 3 * 4 = 4$ 18. $-2 + 3 * 4 = 10$

19. $-3 + 5 * 2 - (-6) = 37$ 20. $4^2 + (-3) - (-5) * 2 = 20$

21. a. Julie arrived 20 minutes before the race began. She started right on time. It took her 24 minutes to finish the 6-kilometer race. She stayed 10 minutes after the race to cool off; then she left. If she arrived at the race at 9:10 A.M., what time was it when she left?

_____

b. Explain how you found your answer.

_____

_____

_____

_____

163

Unit 7 Review

SRB
5-9

1. Circle the number sentences that are true.

$25 + (-6) < -32$ $\qquad$ $4^2 < 2^4$ $\qquad$ $15 * 15 * 15 < 15^3$

$21 * 21 = 21^3$ $\qquad$ $-5 - (-58) = 53$ $\qquad$ $25 > 5^2 - (-2)$

Write each number as a power of 10.

2. 1,000,000 _____

3. 10,000 _____

4. 1 hundred-thousand _____

5. 1 billion _____

Match the number written in number-and-word notation with its standard notation. Fill in the oval next to the correct answer.

6. 3 million

    0 300,000

    0 30,000,000

    0 3,000,000

    0 30,000

7. 20 thousand

    0 200,000

    0 20,000

    0 2,000,000

    0 20,000,000

8. 640 thousand

    0 6,400,000

    0 64,000,000

    0 640,000,000

    0 640,000

9. 2.6 million

    0 26,000,000

    0 2,060,000

    0 20,600,000

    0 2,600,000

Write the following numbers in expanded notation.

10. 8,759 _____

11. 87.59 _____

STUDY LINK 7·11 | **Unit 7 Review** *continued*

SRB
5–9

Write each number in scientific notation.

12. 8 million _____

13. 7 billion _____

14. 3 thousand _____

15. 17 billion _____

16. Louise bought three 6-pack containers of yogurt. She ate 5 individual containers of yogurt in one week. How many containers did she have left?

Number model: _____   Answer: _____

17. The water in Leroy's and Jerod's fish tank had evaporated so it was about $\frac{5}{8}$ inch below the level it should be. They added water and the water level went up about $\frac{3}{4}$ inch. Did the water level end up above or below where it should be?

_____

How much above or below?

Number model: _____   Answer: _____

Find the number that each variable represents.

18. $2.4 + 62.8 + 3.752 = f$ _____

19. $86.54 + b = 87$ _____

20. $33\frac{1}{3}\% + p = 100\%$ _____

21. $6,284 \div 4 = a$ _____

22. $8,463 \div 8 = v$ _____

23. $963 \div 7 = k$ _____

STUDY LINK 7·12

Unit 8: Family Letter

Fractions and Ratios

In Unit 4, your child reviewed equivalent fractions. In this unit, we will apply this knowledge to compute with fractions and mixed numbers. Students will learn that the key to fraction computation with unlike denominators is to find common denominators.

Unit 8 also introduces fraction multiplication. Students will use folded paper to represent fractions of a whole. Then the class will study fraction multiplication using area models, which are diagrams that show a *whole* divided into parts. This concept building will lead to a rule for multiplying fractions:

$$\frac{a}{b} * \frac{c}{d} = \frac{a * c}{b * d}$$

Example: $\frac{2}{5} * \frac{3}{4} = \frac{2 * 3}{5 * 4} = \frac{6}{20}$, or $\frac{3}{10}$

For mixed-number multiplication, students will rename the mixed numbers as fractions, then use the rule to multiply. Finally they rename the product as a mixed number.

Example: $2\frac{1}{2} * 1\frac{2}{3} = \frac{5}{2} * \frac{5}{3} = \frac{5 * 5}{2 * 3} = \frac{25}{6} = 4\frac{1}{6}$

Your child might want to use partial products to solve this problem:

$2\frac{1}{2} * 1\frac{2}{3}$ can be thought of as $(2 + \frac{1}{2}) * (1 + \frac{2}{3})$. There are 4 partial products, as indicated by arrows:

$$2 * 1 = 2$$

$$2 * \frac{2}{3} = \frac{4}{3}$$

$$(2 + \frac{1}{2}) * (1 + \frac{2}{3}) \qquad \frac{1}{2} * 1 = \frac{1}{2}$$

$$\frac{1}{2} * \frac{2}{3} = \frac{2}{6}$$

Add the partial products: $2 + \frac{4}{3} + \frac{1}{2} + \frac{2}{6} = 2 + \frac{8}{6} + \frac{3}{6} + \frac{2}{6} = 2 + \frac{13}{6} = 4\frac{1}{6}$

Your child will play several games such as, *Build-It* and *Fraction Action, Fraction Friction,* to practice sorting fractions and adding fractions with unlike denominators.

Finally, as part of the American Tour, students will explore data related to population distribution and household sizes.

Please keep this Family Letter for reference as your child works through Unit 8.

Vocabulary

Important terms in Unit 8:

area model A model for multiplication problems in which the length and width of a rectangle represent the factors and the area represents the product.

discount The amount by which a price of an item is reduced in a sale, usually given as a fraction or percent of the original price, or as a "percent off." For example, a $4 item on sale for $3 is discounted to 75% or $\frac{3}{4}$ of its original price. A $10.00 item at 10% off costs $9.00, or $\frac{1}{10}$ less than the usual price.

majority A number or amount that is more than half of a total number or amount.

quick common denominator The product of the denominators of two or more fractions. For example, the quick common denominator of $\frac{3}{4}$ and $\frac{5}{6}$ is 4 * 6 = 24. In general, the quick common denominator of $\frac{a}{b}$ and $\frac{c}{d}$ is b * d.

unit fraction A fraction whose numerator is 1. For example, $\frac{1}{2}$, $\frac{1}{3}$, $\frac{1}{8}$, and $\frac{1}{20}$ are unit fractions. Unit fractions are especially useful in converting between measurement systems. For example, because 1 foot = 12 inches you can multiply a number of inches by $\frac{1}{12}$ to convert to feet.

unit percent One percent (1%).

Building Skills through Games

In Unit 8, your child will practice skills with fractions and other numbers by playing the following games. For detailed instructions of most games, see the *Student Reference Book*.

Build-It See *Student Reference Book,* p. 300. This game for partners requires a deck of 16 *Build-It* fraction cards. This game provides practice in comparing and ordering fractions.

Factor Captor See *Student Reference Book,* p. 306. Partners play this game with a calculator and paper and pencil. This game provides practice finding factors of a number.

Mixed-Number Spin See *Student Reference Book,* p. 322. Partners use a spinner to randomly select fractions and mixed numbers, used to complete number sentences. This game provides practice in adding and subtracting fractions and mixed numbers.

Frac-Tac-Toe See *Student Reference Book,* p. 274–276. This game for partners requires a deck of number cards 0–10 and a gameboard similar to a bingo card. The game provides practice converting between fractions, decimals, and percents.

Fraction Action, Fraction Friction See *Student Reference Book,* p. 312. This game for partners requires a set of 16 *Fraction Action, Fraction Friction* cards. The game provides practice adding fractions with unlike denominators.

Name That Number See *Student Reference Book,* p. 325. Partners play a card game. This game provides practice in using order of operations to write number sentences.

Do-Anytime Activities

To work with your child on the key concepts, try these rewarding activities.

1. Ask your child to measure the lengths of two objects using a ruler. Then ask him or her to calculate the sum and difference of their lengths.

2. Ask your child to explain how to use the fraction operation keys on his or her calculator. For example, ask your child to show you how to enter fractions and mixed numbers, simplify fractions, and convert between fractions and decimals.

3. Help your child identify advertisements in signs, newspapers, and magazines that use percents. Help your child find the sale price of an item that is discounted by a certain percent. For example, a $40 shirt reduced by 25% costs $30.

As You Help Your Child with Homework

As your child brings assignments home, you might want to go over the instructions together, clarifying them as necessary. The answers listed below will guide you through this unit's Study Links.

Study Link 8·1

1. $\frac{3}{6}$ **2.** $\frac{2}{3}$ **3.** $\frac{5}{6}$

4. $\frac{19}{20}$ **5.** $\frac{9}{17}$ **6.** $\frac{4}{7}$

7. Sample answer: The quick common denominator is $21 * 17$, or 357. $\frac{11}{21} = \frac{11 * 17}{21 * 17} = \frac{187}{357}$, and $\frac{9}{17} = \frac{9 * 21}{17 * 21} = \frac{189}{357}$. So $\frac{9}{17}$ is greater.

8. 0.75   **9.** $0.\overline{6}$ **10.** 0.625

11. 0.7   **12.** 0.55   **13.** 0.84

14. Sample answer: $\frac{1}{8}$ is half of $\frac{1}{4}$ $\left(\frac{0.25}{2} = 0.125\right)$. $\frac{5}{8} = \frac{4}{8} + \frac{1}{8} = 0.5 + 0.125$, or 0.625.

15. >   **16.** =   **17.** >

18. >   **19.** >   **20.** >

21. Sample answer: $\frac{6}{7} + \frac{1}{7} = 1$. $\frac{1}{8}$ is less than $\frac{1}{7}$, so $\frac{6}{7} + \frac{1}{8}$ is less than 1.

Study Link 8·2

2. 2   **3.** $10\frac{2}{3}$ **5.** $5\frac{1}{2}$

7. 6   **9.** 14   **11.** $5\frac{1}{4}$

13. $9\frac{3}{8}$ **15.** $8\frac{1}{4}$

Study Link 8·3

1. 11   **3.** 10   **6.** $6\frac{5}{3}$

7. $2\frac{1}{2}$ **9.** $2\frac{1}{5}$ **11.** $5\frac{4}{9}$

13. $2\frac{1}{4}$ **15.** $\frac{1}{2}$

Study Link 8·4

1. $\frac{4}{5}$; $\frac{155}{200}$ **2.** $< \frac{1}{2}$ **3.** $> \frac{1}{2}$

4. $= \frac{1}{2}$ **5.** $< \frac{1}{2}$

6. $\frac{\boxed{6}}{\boxed{1}} + \frac{\langle 5 \rangle}{\boxed{6}} = \frac{41}{6} = 6\frac{5}{6}$

Study Link 8·5

1. $\frac{3}{12}$, or $\frac{1}{4}$ **2.** $\frac{6}{15}$, or $\frac{2}{5}$

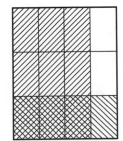 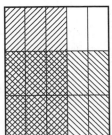

5. Nina: $\frac{1}{2}$; Phillip: $\frac{1}{6}$; Ezra: $\frac{1}{6}$; Benjamin: $\frac{1}{6}$

Study Link 8·6

1. $\frac{1}{3} * \frac{2}{5} = \frac{2}{15}$

3. $\frac{7}{8} * \frac{1}{3} = \frac{7}{24}$

5. $\frac{10}{18}$, or $\frac{5}{9}$

7. $\frac{12}{25}$

9. $\frac{5}{63}$

11. 9; 3

Study Link 8·7

7.

Rule	in (□)	out (△)
$\triangle = \square * 4$	$\frac{2}{3}$	$\frac{8}{3}$, or $2\frac{2}{3}$
	$\frac{4}{5}$	$\frac{16}{5}$, or $3\frac{1}{5}$
	$\frac{8}{9}$	$\frac{32}{9}$, or $3\frac{5}{9}$
	$\frac{5}{4}$	$\frac{20}{4}$, or 5
	$\frac{7}{3}$	$\frac{28}{3}$, or $9\frac{1}{3}$

8.

Rule	in (□)	out (△)
$\triangle = \square * \frac{1}{4}$	2	$\frac{1}{2}$
	3	$\frac{3}{4}$
	$\frac{5}{6}$	$\frac{5}{24}$
	$\frac{2}{3}$	$\frac{1}{6}$

Study Link 8·8

1. a. $\frac{46}{24}$, or $1\frac{11}{12}$ **b.** $\frac{10}{40}$, or $\frac{1}{4}$

c. $\frac{85}{24}$, or $3\frac{13}{24}$ **d.** $\frac{175}{24}$, or $7\frac{7}{24}$

e. $\frac{296}{60}$, or $4\frac{14}{15}$ **f.** $\frac{364}{40}$, or $9\frac{1}{10}$

2. a. $8\frac{5}{9}$ **b.** $5\frac{1}{2}$ **c.** $2\frac{1}{12}$

3. a. 5   **b.** $5\frac{5}{8}$

Study Link 8·9

1. $\frac{45}{100}$; 0.45; 45%

$\frac{3}{10}$; 0.3; 30%

$\frac{2}{10}$; 0.2; 20%

$\frac{15}{100}$; 0.15; 15%

2. Calculated discounts: $100.00; $1,600.00;
$7.84; $0.75; $8.70; $5.28; $810.00; $385.00

Study Link 8·10

1. 4;20   **3.** 1,200 miles

5. 16 min.   **6.** yes

Study Link 8·11

Sample answers for Problems 1–4:

1. $\frac{14}{16}$, $\frac{28}{32}$, $\frac{35}{40}$ **2.** $\frac{6}{8}$, $\frac{9}{12}$, $\frac{12}{16}$

3. $\frac{1}{2}$, $\frac{2}{4}$, $\frac{3}{6}$ **4.** $\frac{4}{6}$, $\frac{6}{9}$, $\frac{8}{12}$

5. $\frac{3}{8}$ **6.** $\frac{5}{9}$

7. $\frac{7}{9}$ **8.** $\frac{7}{12}$

9. Sample answer: I changed $\frac{4}{10}$ and $\frac{7}{12}$ to fractions with a common denominator.
$\frac{4}{10} = \frac{24}{60}$ and $\frac{7}{12} = \frac{35}{60}$. Because $\frac{1}{2} = \frac{30}{60}$, $\frac{7}{12}$ is $\frac{5}{60}$ away from $\frac{1}{2}$, and $\frac{4}{10}$ is $\frac{6}{60}$ away from $\frac{1}{2}$. So, $\frac{7}{12}$ is closer to $\frac{1}{2}$.

11. $\frac{11}{18}$ **13.** $\frac{17}{24}$

14. $\frac{3}{10}$ **15.** $3\frac{1}{3}$

Study Link 8·12

1. 5   **2.** 22

3. $3\frac{4}{5}$ **5.** $1\frac{5}{9}$

7. $8\frac{5}{12}$ **9.** $11\frac{1}{4}$

11. $4\frac{1}{2}$ **13.** $\frac{15}{2}$, or $7\frac{1}{2}$

STUDY LINK
8·1   **Comparing Fractions**

SRB
66–68
83–88

Circle the greater fraction for each pair.

1. $\frac{3}{8}$ or $\frac{3}{6}$

2. $\frac{2}{3}$ or $\frac{2}{9}$

3. $\frac{4}{7}$ or $\frac{5}{6}$

4. $\frac{19}{20}$ or $\frac{4}{8}$

5. $\frac{11}{21}$ or $\frac{9}{17}$

6. $\frac{4}{7}$ or $\frac{6}{11}$

7. Explain how you got your answer for Problem 5.

_____

_____

_____

Write the decimal equivalent for each fraction.

8. $\frac{3}{4}$ = _____

9. $\frac{2}{3}$ = _____

10. $\frac{5}{8}$ = _____

11. $\frac{7}{10}$ = _____

12. $\frac{11}{20}$ = _____

13. $\frac{21}{25}$ = _____

14. Explain how you can do Problem 10 without using a calculator.

_____

_____

Use >, <, or = to make each number sentence true.

15. $\frac{1}{2} + \frac{5}{8}$ _____ 1

16. $\frac{2}{3} + \frac{2}{6}$ _____ 1

17. $\frac{7}{9} + \frac{3}{5}$ _____ 1

18. 1 _____ $\frac{6}{10} + \frac{5}{20}$

19. 1 _____ $\frac{3}{8} + \frac{4}{9}$

20. 1 _____ $\frac{6}{7} + \frac{1}{8}$

21. Explain how you found the answer to Problem 20.

_____

_____

Practice

22. 675 * 42 = _____

23. 28,350 ÷ 675 = _____

24. 67.5 − 0.42 = _____

25. 28,350 + 42 + 67.08 = _____

STUDY LINK
8·2

Adding Mixed Numbers

Rename each mixed number in simplest form.

1. $3\frac{6}{5} = $ $4\frac{1}{5}$

2. $\frac{16}{8} = $ _____

3. $9\frac{5}{3} = $ _____

4. $1\frac{7}{5} = $ _____

5. $4\frac{6}{4} = $ _____

6. $5\frac{10}{6} = $ _____

Add. Write each sum as a whole number or mixed number in simplest form.

7. $3\frac{1}{4} + 2\frac{3}{4} = $ _____

8. $4\frac{1}{5} + 3\frac{4}{5} = $ _____

9. $9\frac{1}{3} + 4\frac{2}{3} = $ _____

10. $3\frac{5}{7} + 8\frac{6}{7} = $ _____

11. $\frac{15}{8} + 3\frac{3}{8} = $ _____

12. $4\frac{2}{9} + 5\frac{5}{9} = $ _____

Add.

13. $2\frac{5}{8}$
    $+ 6\frac{3}{4}$
    _____

14. $7\frac{1}{2}$
    $+ 3\frac{2}{3}$
    _____

15. $4\frac{6}{9}$
    $+ 3\frac{7}{12}$
    _____

16. $5\frac{3}{4}$
    $+ 2\frac{4}{5}$
    _____

Practice

17. $3,540 \div 6 = $ _____

18. $1,770 \div 3 = $ _____

19. $7,080 / 12 = $ _____

20. $(590 * 5) \div 2 = $ _____

STUDY LINK 8·3 | **Subtracting Mixed Numbers**

Fill in the missing numbers.

1. $3\frac{3}{8} = 2\frac{\square}{8}$

2. $4\frac{5}{6} = \square\frac{11}{6}$

3. $2\frac{1}{9} = 1\frac{\square}{9}$

4. $6\frac{3}{7} = \square\frac{10}{7}$

5. $4\frac{3}{5} = 3\frac{\square}{5}$

6. $7\frac{2}{3} = \square\frac{\square}{3}$

Subtract. Write your answers in simplest form.

7. $\begin{array}{r} 5\frac{3}{4} \\ -\ 3\frac{1}{4} \\ \hline \end{array}$

8. $\begin{array}{r} 6\frac{2}{3} \\ -\ 4\frac{1}{3} \\ \hline \end{array}$

9. $\begin{array}{r} 5\frac{4}{5} \\ -\ 3\frac{3}{5} \\ \hline \end{array}$

10. $4 - \frac{3}{8} =$ _____

11. $6 - \frac{5}{9} =$ _____

12. $5 - 2\frac{3}{10} =$ _____

13. $7 - 4\frac{3}{4} =$ _____

14. $3\frac{2}{5} - 1\frac{3}{5} =$ _____

15. $4\frac{3}{8} - 3\frac{7}{8} =$ _____

Practice

16. $654 * 205 =$ _____

17. $654 * 502 =$ _____

18. $654 * 250 =$ _____

19. $654 * 520 =$ _____

STUDY LINK 8·4 | **More Fraction Problems**

1. Circle all the fractions below that are greater than $\frac{3}{4}$.

$\frac{4}{5}$ $\frac{13}{20}$ $\frac{1}{2}$ $\frac{18}{25}$ $\frac{9}{12}$ $\frac{155}{200}$ $\frac{7}{11}$

Rewrite each expression by renaming the fractions with a common denominator. Then decide whether the sum or difference is greater than $\frac{1}{2}$, less than $\frac{1}{2}$, or equal to $\frac{1}{2}$. Circle your answer.

2. $\frac{1}{10} + \frac{2}{7}$ _____ $> \frac{1}{2}$ $< \frac{1}{2}$ $= \frac{1}{2}$

3. $\frac{5}{6} - \frac{1}{4}$ _____ $> \frac{1}{2}$ $< \frac{1}{2}$ $= \frac{1}{2}$

4. $\frac{18}{20} - \frac{2}{5}$ _____ $> \frac{1}{2}$ $< \frac{1}{2}$ $= \frac{1}{2}$

5. $\frac{3}{4} - \frac{1}{3}$ _____ $> \frac{1}{2}$ $< \frac{1}{2}$ $= \frac{1}{2}$

Fraction Puzzle

6. Select and place three different numbers so the sum is as large as possible.

Procedure: Select three different numbers from this list: 1, 2, 3, 4, 5, 6.

- ◆ Write the same number in each square.
- ◆ Write a different number in the circle.
- ◆ Write a third number in the hexagon.
- ◆ Add the two fractions.

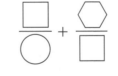

Example: $\dfrac{\boxed{2}}{\bigcirc 4} + \dfrac{\langle 3 \rangle}{\boxed{2}} = \dfrac{8}{4} = 2$

Practice

7. $3 - 2.564 =$ _____   8. $3 * 2.564 =$ _____

9. $16 - 5.438 =$ _____   10. $3,049 / 15 =$ _____

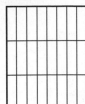

Fractions of a Fraction

Example:

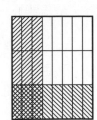

The whole rectangle represents ONE.

Shade $\frac{3}{8}$ of the interior.

Shade $\frac{1}{3}$ of the interior in a different way.

The double shading shows that $\frac{1}{3}$ of $\frac{3}{8}$ is $\frac{3}{24}$, or $\frac{1}{8}$.

In each of the following problems, the whole rectangle represents ONE.

1. Shade $\frac{3}{4}$ of the interior.

Shade $\frac{1}{3}$ of the interior in a different way.

The double shading shows that

$\frac{1}{3}$ of $\frac{3}{4}$ is _____.

2. Shade $\frac{3}{5}$ of the interior.

Shade $\frac{2}{3}$ of the interior in a different way.

The double shading shows that

$\frac{2}{3}$ of $\frac{3}{5}$ is _____.

3. Shade $\frac{4}{5}$.

Shade $\frac{3}{4}$ of the interior in a different way.

The double shading shows that

$\frac{3}{4}$ of $\frac{4}{5}$ is _____.

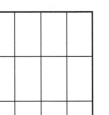

4. Shade $\frac{5}{8}$.

Shade $\frac{3}{5}$ of the interior in a different way.

The double shading shows that

$\frac{3}{5}$ of $\frac{5}{8}$ is _____.

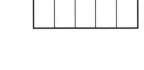

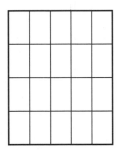

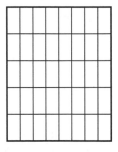

5. Nina and Phillip cut Mr. Ferguson's lawn. Nina worked alone on her half, but Phillip shared his half equally with his friends, Ezra and Benjamin. What fraction of the earnings should each person get?

179

**STUDY LINK
8·6**

Multiplying Fractions

Write a number model for each area model.

Example:

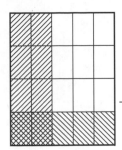

$$\frac{1}{4} * \frac{2}{5} = \frac{2}{20}, \text{ or } \frac{1}{10}$$

1.

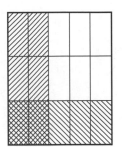

_____

2.

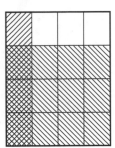

_____

3.

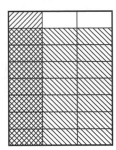

_____

> **Reminder:** $\frac{a}{b} * \frac{c}{d} = \frac{a*c}{b*d}$

Multiply.

4. $\frac{3}{7} * \frac{2}{10} =$ _____

5. $\frac{5}{6} * \frac{2}{3} =$ _____

6. $\frac{1}{2} * \frac{1}{4} =$ _____

7. $\frac{4}{5} * \frac{3}{5} =$ _____

8. $\frac{2}{3} * \frac{3}{8} =$ _____

9. $\frac{1}{7} * \frac{5}{9} =$ _____

10. Matt is making cookies for the school fund-raiser. The recipe calls for $\frac{2}{3}$ cup of chocolate chips. He decides to triple the recipe. How many cups of chocolate chips does he need? _____ cups

11. The total number of goals scored by both teams in the field-hockey game was 15. Julie's team scored $\frac{3}{5}$ of the goals. Julie scored $\frac{1}{3}$ of her team's goals. How many goals did Julie's team score? _____ goals

How many goals did Julie score? _____ goals

181

Multiplying Fractions and Whole Numbers

Use the fraction multiplication algorithm to calculate the following products.

SRB
73

1. $\frac{5}{3} * 9 =$ _____

2. $\frac{3}{8} * 12 =$ _____

3. $\frac{1}{8} * 5 =$ _____

4. $20 * \frac{3}{4} =$ _____

5. $\frac{5}{6} * 14 =$ _____

6. $27 * \frac{2}{9} =$ _____

7. Use the given rule to complete the table.

Rule	
$\triangle = \square * 4$	

in ($\square$)	out ($\triangle$)
$\frac{2}{3}$	
$\frac{4}{5}$	
$\frac{8}{9}$	
$\frac{5}{4}$	
$\frac{7}{3}$	

8. What is the rule for the table below?

Rule	

in ($\square$)	out ($\triangle$)
2	$\frac{1}{2}$
3	$\frac{3}{4}$
$\frac{5}{6}$	$\frac{5}{24}$
$\frac{2}{3}$	$\frac{1}{6}$

9. Make and complete your own "What's My Rule?" table on the back of this page.

183

STUDY LINK 8·9

Fractions, Decimals, and Percents

1. Complete the table so each number is shown as a fraction, decimal, and percent.

Fraction	Decimal	Percent
		45%
	0.3	
$\frac{2}{10}$		
	0.15	

2. Use your percent sense to estimate the discount for each item. Then calculate the discount for each item. (If necessary, round to the nearest cent.)

Item	List Price	Percent of Discount	Estimated Discount	Calculated Discount
Saguaro cactus with arms	$400.00	25%		
Life-size wax figure of yourself	$10,000.00	16%		
Manhole cover	$78.35	10%		
Live scorpion	$14.98	5%		
10,000 honeybees	$29.00	30%		
Dinner for one on the Eiffel Tower	$88.00	6%		
Magician's box for sawing a person in half	$4,500.00	18%		
Fire hydrant	$1,100.00	35%		

Source: *Everything Has Its Price*

187

Unit Fractions

SRB
74 75

Finding the worth of the unit fraction will help you solve each problem below.

1. If $\frac{4}{5}$ of a number is 16, what is $\frac{1}{5}$ of the number? _____

   What is the number? _____

2. Our football team won $\frac{3}{4}$ of the games that it played.
   It won 12 games. How many games did it play?
   _____
   (unit)

3. When a balloon had traveled 800 miles, it had completed
   $\frac{2}{3}$ of its journey. What was the total length of its trip?
   _____
   (unit)

4. Grandpa baked cookies. Twenty cookies were oatmeal
   raisin. The oatmeal raisin cookies represent $\frac{5}{8}$ of all
   the cookies. How many cookies did Grandpa bake?
   _____
   (unit)

5. Tiana jogged $\frac{6}{8}$ of the way to school in 12 minutes. If
   she continues at the same speed, how long will her entire
   jog to school take?
   _____
   (unit)

6. After 35 minutes, Hayden had completed $\frac{7}{10}$ of his math test.
   If he has a total of 55 minutes to complete the test, do you
   think he will finish in time?
   _____

   Explain: _____

   _____

7. Complete the table using the given rule.

Rule
out = 60% of in

in	out
100	
60	
	42
110	
	72
35	

8. Find the rule. Then complete the table.

Rule
out = _____ of in

in	out
24	9
72	27
56	21
80	30
	15
32	

189

STUDY LINK 8·11 | **Fraction Review**

Write three equivalent fractions for each fraction.

1. $\frac{7}{8}$ _____

2. $\frac{3}{4}$ _____

3. $\frac{6}{12}$ _____

4. $\frac{2}{3}$ _____

Circle the fraction that is closer to $\frac{1}{2}$.

5. $\frac{3}{8}$ or $\frac{4}{5}$ 6. $\frac{4}{7}$ or $\frac{5}{9}$ 7. $\frac{7}{8}$ or $\frac{7}{9}$ 8. $\frac{4}{10}$ or $\frac{7}{12}$

9. Explain how you found your answer for Problem 8.

_____

_____

_____

_____

Solve. Write your answers in simplest form.

10. _____ $= \frac{5}{6} + \frac{3}{4}$

11. $\frac{7}{9} - \frac{1}{6} =$ _____

12. $8 - \frac{2}{3} =$ _____

13. $\frac{7}{8} - \frac{1}{6} =$ _____

14. $\frac{3}{4}$ of $\frac{2}{5}$ is _____.

15. $4 * \frac{5}{6} =$ _____

Practice

16. $64,072 - 15,978 =$ _____

17. $2,297 \div 45 \rightarrow$ _____

18. $1,674 - 1,204 =$ _____

19. $326 + 684 + 934 =$ _____

191

Mixed-Number Review

Fill in the missing numbers.

1. $4\frac{1}{4} = 3\frac{\boxed{}}{4}$

2. $\dfrac{\boxed{}}{5} = 3\frac{7}{5}$

Solve. Write your answers in simplest form.

3. $1\frac{3}{5} + 2\frac{1}{5} = $ _____

4. $3\frac{3}{8} - 1\frac{5}{8} = $ _____

5. $7\frac{4}{9} - 5\frac{8}{9} = $ _____

6. $3\frac{2}{7} + 1\frac{4}{5} = $ _____

7. $5\frac{2}{3} + 2\frac{3}{4} = $ _____

8. $4 - 1\frac{3}{4} = $ _____

9. $3 * 3\frac{3}{4} = $ _____

10. $4\frac{2}{3} * \frac{6}{7} = $ _____

11. _____ $= 2\frac{1}{2} * 1\frac{4}{5}$

12. $\frac{3}{10} * 8\frac{1}{3} = $ _____

Common Denominator Division

Here is one way to divide fractions and to divide whole or mixed numbers by fractions.

Step 1 Rename the numbers using a common denominator.

Step 2 Divide the numerators.

Solve. Show your work.

13. $5 \div \frac{2}{3} = $ _____

14. $\frac{4}{7} \div \frac{3}{5} = $ _____

15. $4\frac{1}{8} \div \frac{3}{4} = $ _____

16. $6\frac{2}{3} \div \frac{7}{9} = $ _____

193

STUDY LINK 8·13 | Unit 9: Family Letter

Coordinates, Area, Volume, and Capacity

In the beginning of this unit, your child will practice naming and locating ordered number pairs on a coordinate grid. Whole numbers, fractions, and negative numbers will be used as coordinates. Your child will play the game *Hidden Treasure,* which provides additional practice with coordinates. You might want to challenge your child to a round.

In previous grades, your child studied the perimeters (distances around) and the areas (amounts of surface) of geometric figures. *Fourth Grade Everyday Mathematics* developed and applied formulas for the areas of rectangles, parallelograms, and triangles. In this unit, your child will review these formulas and explore new area topics, including the rectangle method for finding areas of regular and irregular shapes.

Students will also examine how mathematical transformations change the area, perimeter, and angle measurements of a figure. These transformations resemble changes and motions in the physical world. In some transformations, figures are enlarged in one or two dimensions; in other transformations, figures are translated (slid) or reflected (flipped over).

In the Earth's Water Surface exploration, students locate places on Earth with latitude and longitude. Then they use latitude and longitude in a sampling experiment that enables them to estimate, without measuring, the percent of Earth's surface that is covered by water. In the School's Land Area exploration, students use actual measurements and scale drawings to estimate their school's land area.

The unit concludes with a look at volume (the amount of space an object takes up) and capacity (the amount of material a container can hold). Students develop a formula for the volume of a prism (volume = area of the base * the height). They observe the metric equivalents 1 liter = 1,000 milliliters = 1,000 cubic centimeters, and they practice making conversions between U.S. customary measures (1 gallon = 4 quarts, and so on).

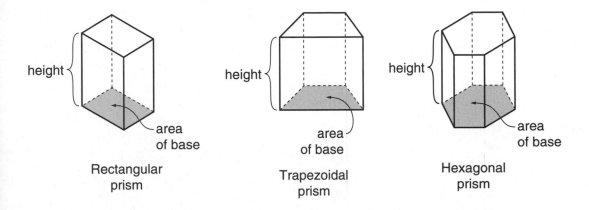

height { area of base

Rectangular prism

height { area of base

Trapezoidal prism

height { area of base

Hexagonal prism

Please keep this Family Letter for reference as your child works through Unit 9.

195

Vocabulary

Important terms in Unit 9:

area The amount of surface inside a 2-dimensional figure. Area is measured in square units, such as square inches (in^2) and square centimeters (cm^2).

axis of a coordinate grid Either of the two number lines that intersect to form a coordinate grid.

capacity The amount of space occupied by a *3-dimensional* shape. Same as *volume*. The amount a container can hold. Capacity is often measured in units such as *quarts, gallons, cups,* or *liters.*

coordinate A number used to locate a point on a number line, or one of two numbers used to locate a point on a coordinate grid.

coordinate grid A reference frame for locating points in a plane using ordered number pairs, or coordinates.

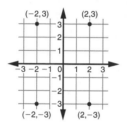

Rectangular coordinate grid

formula A general rule for finding the value of something. A formula is usually an equation with quantities represented by letter *variables*. For example, the formula for the area of a rectangle may be written as $A = \ell * w$, where A represents the area of the rectangle, ℓ represents the length, and w represents the width.

latitude A measure, in degrees, of the distance of a place north or south of the equator.

longitude A measure, in degrees, of how far east or west of the prime meridian a place is.

ordered number pair Two numbers that are used to locate a point on a *coordinate grid.* The first number gives the position along the horizontal axis; the second number gives the position along the vertical axis. Ordered number pairs are usually written inside parentheses: (2,3).

perpendicular Two lines or two planes that intersect at right angles. Line segments or rays that lie on perpendicular lines are perpendicular to each other. The symbol ⊥ means *is perpendicular to.*

rectangle method A method for finding area in which one or more rectangles are drawn around a figure or parts of a figure.

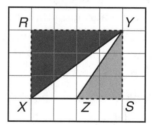

To find the area of triangle *XYZ,* first draw rectangle *XRYS* through its vertices. Then subtract the areas of the two shaded triangles from the area of rectangle *XRYS.*

transformation Something done to a geometric figure that produces a new figure. Common transformations are translations (slides), reflections (flips), and rotations (turns).

volume The amount of space occupied by a 3-dimensional shape. Same as *capacity.* The amount a container can hold. Volume is usually measured in cubic units, such as cubic centimeters (cm^3), cubic inches (in^3), or cubic feet (ft^3).

Do-Anytime Activities

To work with your child on concepts taught in this unit, try these interesting and rewarding activities:

1. Find an atlas or map that uses letter-number pairs to locate places. For example, an atlas might say that Chattanooga, Tennessee, is located at D-9. Use the letter-number pairs to locate places you have visited or would like to visit.

2. Estimate the area of a room in your home. Use a tape measure or ruler to measure the room's length and width, and multiply to find the area. Make a simple sketch of the room, including the length, the width, and the area. If you can, find the area of other rooms or of your entire home.

Building Skills through Games

In Unit 9, your child will develop his or her understanding of coordinates and coordinate grids by playing the following games. For detailed instructions, see the *Student Reference Book.*

Frac-Tac-Toe See *Student Reference Book,* pages 309–311. Two players use a set of number cards 0–10 (4 of each), a gameboard, counters, and a calculator to play one of many versions. Students practice converting between fractions, decimals, and percents.

Hidden Treasure See *Student Reference Book,* page 319. This game for 2 players provides practice using coordinates and coordinate grids. It also offers the opportunity for players to develop good search strategies. Each player will need a pencil and two 1-quadrant playing grids with axes labeled from 0 to 10.

Polygon Capture See *Student Reference Book,* page 328. This game involves two to four players. Materials include polygon pieces and property cards. Players strengthen skills with identifying attributes of polygons. Players may also use 4-quadrant grids with axes labeled from −7 to 7. Practice is extended to coordinates and grids that include negative numbers.

As You Help Your Child with Homework

As your child brings assignments home, you might want to go over the instructions together, clarifying them as necessary. The answers listed below will guide you through some of the Study Links in this unit.

Study Link 9·1

2. Rectangular prism

3. a. (11,7)   **4.** 13,297

5. 872.355   **6.** $10\frac{2}{8}$, or $10\frac{1}{4}$

Study Link 9·2

1. Sample answers: (8,16); (0,5); (16,5)

2. isosceles   **4.** quadrangle

Study Link 9·3

2. The first number

3.

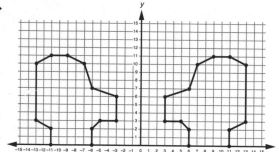

4. 26,320   **6.** $\frac{14}{24}$, or $\frac{7}{12}$

Study Link 9·4

1. 150 sq ft; 12 hr 30 min   **2.** 114 square feet

3. 80 yd^2 **4.** 33 ft^2

5.

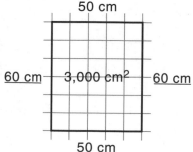

50 cm

60 cm — 3,000 cm^2 — 60 cm

50 cm

6.

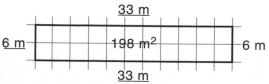

33 m

6 m — 198 m^2 — 6 m

33 m

Study Link 9·5

1. 4 cm^2 **2.** 6 cm^2 **3.** 16 cm^2

4. 10 cm^2 **5.** 15 cm^2 **6.** 4 cm^2

Study Link 9·6

1. 4.5 cm^2; $\frac{1}{2} * 3 * 3 = 4.5$

2. 7.5 cm^2; $\frac{1}{2} * 5 * 3 = 7.5$

3. 3 cm^2; $\frac{1}{2} * 2 * 3 = 3$

4. 24 cm^2; $6 * 4 = 24$

5. 12 cm^2; $4 * 3 = 12$

6. 8 cm^2; $4 * 2 = 8$

Study Link 9·7

1. yd^2 **2.** cm^2 **3.** cm^2

4. in^2 **5.** ft^2

6. $A = \frac{1}{2} * 20 * 13$; 130 ft^2 **7.** $A = 8 * 2$; 16 cm^2

8. $A = \frac{1}{2} * 22 * 7$; 77 yd^2 **9.** $A = 8 * 9\frac{1}{2}$; 76 m^2

Study Link 9·8

1. 15 cm^2; 15 cm^3; 45 cm^3 **2.** 8 cm^2; 8 cm^3; 16 cm^3

3. 9 cm^2; 9 cm^3; 27 cm^3 **4.** 14 cm^2; 14 cm^3; 56 cm^3

5. $\frac{3}{40}$ **6.** 960   **7.** 3,840

Study Link 9·9

1. 72 cm^3 **2.** 144 cm^3 **3.** 70 in^3

4. 162 cm^3 **5.** 45 in^3 **6.** 140 m^3

7. 4   **8.** −245   **9.** 160

Study Link 9·10

2. $A = \frac{1}{2} * 7 * 6$; 21 cm^2 **3.** $A = 8 * 6$; 48 in^2

STUDY LINK 9·1 Plotting Points

1. Plot the following points on the grid below. After you plot each point, draw a line segment to connect it to the last point you plotted.
Reminder: Use your straightedge!

(3,6); (11,11); (15,11); (15,7); (7,2); (3,2); (3,6); (7,6)

Draw a line segment connecting (7,6) and (7,2).
Draw a line segment connecting (7,6) and (15,11).

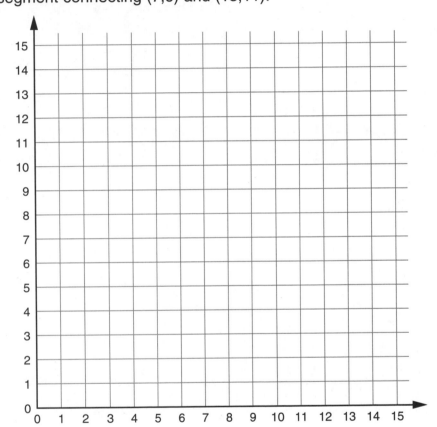

2. What 3-dimensional shape could this drawing represent? _____

3. a. What ordered pair would name the missing vertex to represent a prism? _____

   b. Draw the missing vertex, and then add dashed lines for the missing edges.

Practice

4. $3{,}745 + 8{,}761 + 791 =$ _____

5. $3.745 + 87.61 + 781 =$ _____

6. $4\frac{3}{8} + 5\frac{7}{8} =$ _____

7. $\frac{1}{5} + \frac{3}{4} =$ _____

STUDY LINK 9·2 | Plotting Figures on a Coordinate Grid

1. Plot three points, and make a triangle on the grid below. Label the points as *A*, *B*, and *C*. List the coordinates of the points you've drawn.

A: (_____, _____)   B: (_____, _____)   C: (_____, _____)

2. Circle the name of the kind of triangle you drew.

　　　　　　scalene　　　　equilateral　　　　isosceles

3. Plot four points, and make a parallelogram on the grid below. Label the points as *M, N, O,* and *P*. List the coordinates of the points you've drawn.

M: (_____, _____)　　N: (_____, _____)　　O: (_____, _____)　　P: (_____, _____)

4. Circle another name for the parallelogram you've drawn.

　　　　quadrangle　　　　rhombus　　　　rectangle　　　　square

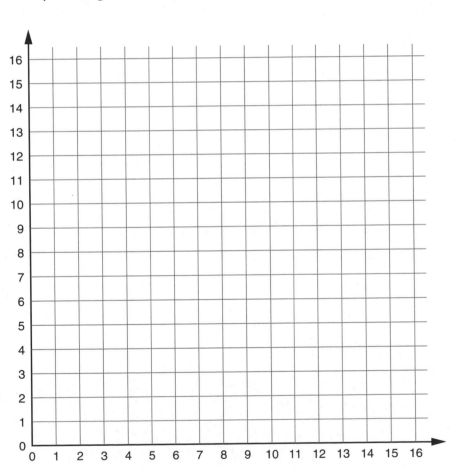

STUDY LINK 9·3 | Reflections on a Coordinate Grid

1. Plot the points listed below. Use a straightedge to connect the points in the same order that you plot them.

   (6,0); (6,2); (5,3); (3,3); (3,6); (6,7); (7,10); (9,11); (11,11); (13,10); (13,3); (11,2); (11,0)

2. Which number (the first number or the second number) in the pair do you need to change to the opposite in order to draw the reflection of this design on the other side of the *y*-axis?

   _____

3. Draw the reflection described above. Plot the points and connect them.

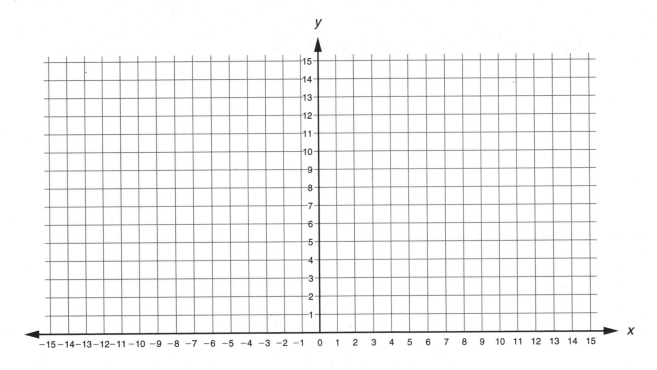

Practice

Multiply.

4. 752 * 35 = _____

5. 75.2 * 0.35 = _____

6. $\frac{7}{8} * \frac{2}{3}$ = _____

7. $2\frac{1}{2} * \frac{3}{4}$ = _____

More Area Problems

SRB
104 105
189

1. Rashid can paint 2 square feet of fence in 10 minutes. Fill in the missing parts to tell how long it will take him to paint a fence that is 6 feet high by 25 feet long. Rashid will be able to paint

_____ of fence in _____.
   (area)   (hours/minutes)

2. Regina wants to cover one wall of her room with wallpaper. The wall is 9 feet high and 15 feet wide. There is a doorway in the wall that is 3 feet wide and 7 feet tall. How many square feet of wallpaper will she need to buy?

_____

Calculate the areas for the figures below.

3.
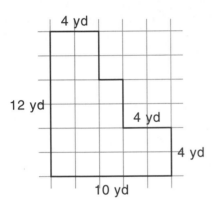

Area = _____ yd²

4.
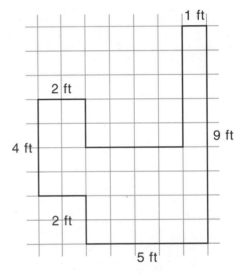

Area = _____ ft²

Fill in the missing lengths for the figures below.

5.
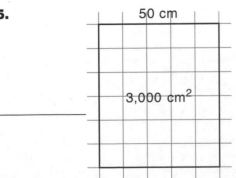

6.
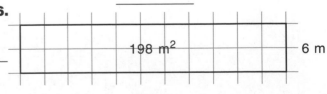

STUDY LINK
9·5

The Rectangle Method

Use the rectangle method to find the area of each figure below.

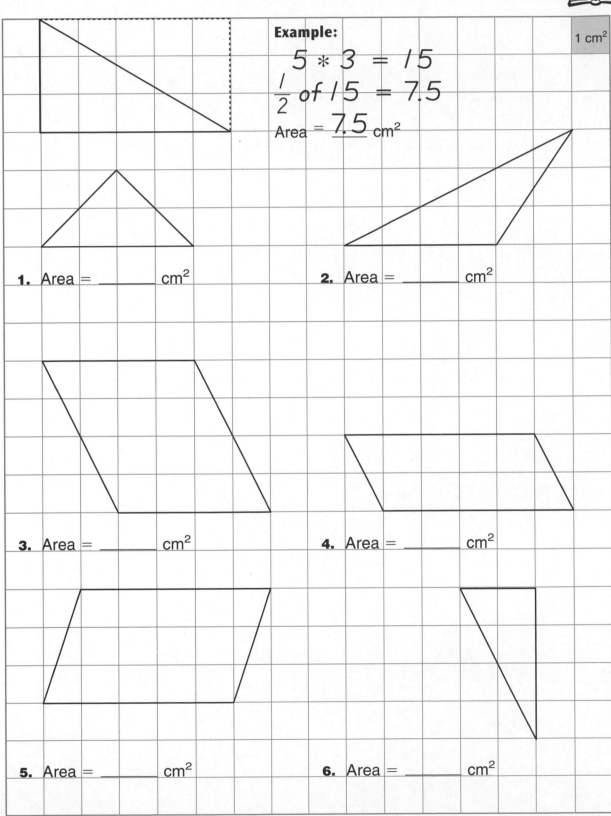

Example:

$5 * 3 = 15$

$\frac{1}{2}$ of $15 = 7.5$

Area = __7.5__ cm²

1 cm²

1. Area = _____ cm²

2. Area = _____ cm²

3. Area = _____ cm²

4. Area = _____ cm²

5. Area = _____ cm²

6. Area = _____ cm²

207

STUDY LINK 9·6 | Area Formulas

For each figure below, label the base and the height, find the area, and record the number model you use to find the area.

> Area of a parallelogram: $A = b * h$
>
> Area of a triangle: $A = \frac{1}{2} * b * h$

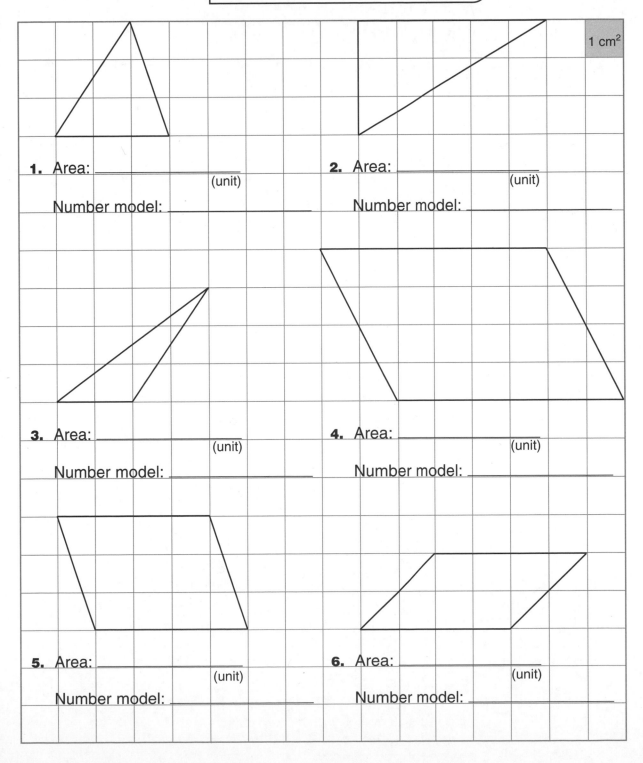

1 cm²

1. Area: _____
   (unit)

   Number model: _____

2. Area: _____
   (unit)

   Number model: _____

3. Area: _____
   (unit)

   Number model: _____

4. Area: _____
   (unit)

   Number model: _____

5. Area: _____
   (unit)

   Number model: _____

6. Area: _____
   (unit)

   Number model: _____

STUDY LINK 9·7 An Area Review

Circle the most appropriate unit to use for measuring the area of each object.

1. The area of a football field

cm²	ft²	yd²	in²

2. The area of your hand

cm²	ft²	yd²	in²

3. The area of a postage stamp

cm²	ft²	yd²	in²

4. Area of a triangular kite

cm²	ft²	yd²	in²

5. Area of a parallelogram-shaped sign on the highway

cm²	ft²	yd²	in²

Use a formula to find the area of each figure. Write the appropriate number sentence and the area.

6.

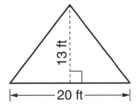

13 ft

20 ft

Number sentence: _____

Area: _____
(unit)

7.

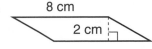

8 cm

2 cm

Number sentence: _____

Area: _____
(unit)

8.

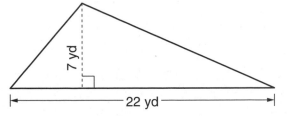

7 yd

22 yd

Number sentence: _____

Area: _____
(unit)

9.

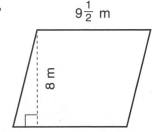

$9\frac{1}{2}$ m

8 m

Number sentence: _____

Area: _____
(unit)

Name	Date	Time

Volumes of Cube Structures

The structures below are made up of centimeter cubes.

1.

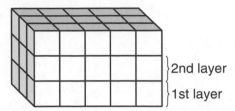

} 2nd layer
} 1st layer

Area of base = _____ cm²

Volume of first layer = _____ cm³

Volume of entire
cube structure = _____ cm³

2.

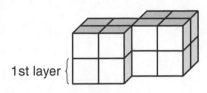

1st layer {

Area of base = _____ cm²

Volume of first layer = _____ cm³

Volume of entire
cube structure = _____ cm³

3.

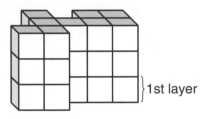

} 1st layer

Area of base = _____ cm²

Volume of first layer = _____ cm³

Volume of entire
cube structure = _____ cm³

4.

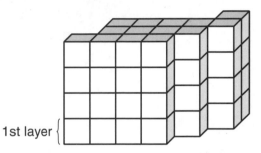

1st layer {

Area of base = _____ cm²

Volume of first layer = _____ cm³

Volume of entire
cube structure = _____ cm³

Practice

5. $\frac{3}{5} * \frac{1}{8} =$ _____

6. $3,840 / 4 =$ _____

7. $960 * 4 =$ _____

8. $\frac{4}{5} * \frac{5}{6} =$ _____

213

STUDY LINK 9·9   **Volumes of Prisms**

The volume *V* of any prism can be found with the formula $V = B * h$, where *B* is the area of the base of the prism, and *h* is the height of the prism from that base.

1.

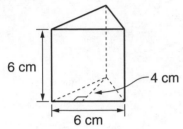

6 cm   4 cm

6 cm

Volume = _____ cm³

2.

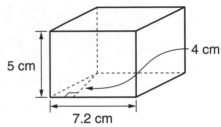

5 cm   4 cm

7.2 cm

Volume = _____ cm³

3.

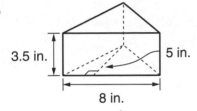

3.5 in.   5 in.

8 in.

Volume = _____ in³

4.

4 cm

6 cm   3 cm

5 cm

3 cm

Volume = _____ cm³

5.

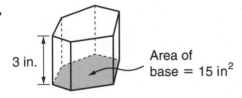

3 in.   Area of base = 15 in²

Volume = _____ in³

6.

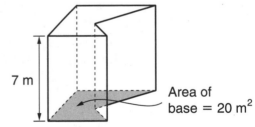

7 m   Area of base = 20 m²

Volume = _____ m³

Practice

Solve each equation.

7. $36 * r = 144$ _____

8. $3,577 - t = 3,822$ _____

9. $3,577 - m = 3,417$ _____

10. $d * 68 = 340$ _____

STUDY LINK
9·10 # Unit 9 Review

1. Plot 6 points on the grid below and connect them to form a hexagon. List the coordinates of the points you plotted.

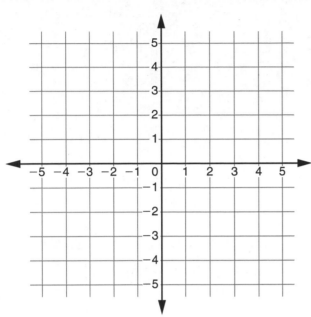

(_____ , _____)

(_____ , _____)

(_____ , _____)

(_____ , _____)

(_____ , _____)

(_____ , _____)

Find the area of the figures shown below. Write the number model you used to find the area.

Area of a rectangle: $A = b * h$

Area of a parallelogram: $A = b * h$

Area of a triangle: $A = \frac{1}{2} * b * h$

2.

7 cm   6 cm

Number model: _____

Area: _____
(unit)

3.

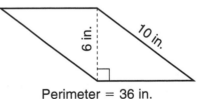

6 in.   10 in.

Perimeter = 36 in.

Number model: _____

Area: _____
(unit)

4. On the back of this page, explain how you solved Problem 3.

217

Unit 10: Family Letter

Algebra Concepts and Skills

In this unit, your child will be introduced to solving simple equations with a pan balance, thus developing basic skills of algebra. For example, a problem might ask how many marbles in the illustration below weigh as much as a cube. You can solve this problem by removing 3 marbles from the left pan and 3 marbles from the right pan. Then the pans will still balance. Therefore, you know that one cube weighs the same as 11 marbles.

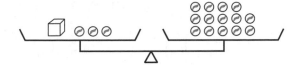

You can think of this pan-balance problem as a model for the equation $c + 3 = 14$, in which the value of c is 11.

A "What's My Rule?" table has been a routine since the early grades of *Everyday Mathematics*. In this unit, your child will follow rules to complete tables, such as the one below and will then graph the data. Your child will also determine rules from information provided in tables and graphs. Students will begin to express such rules using algebraic expressions containing variables.

Rule	in	out
+ 6	−1	5
	2	8
	5	
		12
	12	
		15

As the American Tour continues, your child will work with variables and formulas to predict eruption times of the famous geyser, Old Faithful, in Yellowstone National Park.

In previous grades, your child studied the perimeter (distance around) and the area (amount of surface) of geometric figures. In Unit 9, students developed and applied formulas for the area of rectangles, parallelograms, and triangles. In this unit, your child will explore and apply formulas for the circumference (distance around) and area of circles.

Please keep this Family Letter for reference as your child works through Unit 10.

Vocabulary

Important terms in Unit 10:

algebraic expression An expression that contains a variable. For example, if Maria is 2 inches taller than Joe, and if the variable *M* represents Maria's height, then the algebraic expression $M - 2$ represents Joe's height.

line graph A graph in which data points are connected by line segments.

Attendance for the First Week of School

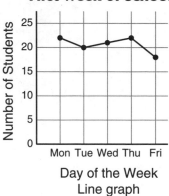

Day of the Week
Line graph

pan balance A tool used to weigh objects or compare weights.

Pan balance

predict In mathematics, to say what will happen in the future based on experimental data or theoretical calculation.

rate A comparison by division of two quantities with unlike units. For example, a speed such as 55 miles per hour is a rate that compares distance with time.

Do-Anytime Activities

To work with your child on concepts taught in this unit and in previous units, try these interesting and rewarding activities:

1. Have your child list different timed distances for a mile. For example, the fastest mile run by a human and by a race car; your child's own fastest mile completed by running, biking, or walking; the fastest mile run for a handicapped athlete; the fastest mile completed by a swimmer, and so on.

2. Have your child keep a running tally of when the school bus arrives. Or have your child time himself or herself to see how long it takes to walk to school in the morning compared to walking home in the afternoon. After a week, have your child describe landmarks for their data and interpret these landmarks.

Building Skills through Games

In this unit, your child will practice using algebraic expressions containing variables by playing the following game. For more detailed instructions, see the *Student Reference Book.*

First to 100 See *Student Reference Book,* page 308.

This is a game for two to four players and requires 32 Problem Cards and a pair of six-sided dice. Players answer questions after substituting numbers for the variable on the Problem Cards. The questions offer practice on a variety of mathematical topics.

As You Help Your Child with Homework

As your child brings assignments home, you might want to go over the instructions together, clarifying them as necessary. The answers listed below will guide you through some of the Study Links in this unit.

Study Link 10·1

1. 3 **2.** 3 **3.** 36 **4.** 4 **5.** 3

Study Link 10·2

3. 5, 10 **4.** 2, 2 **5.** 4, 6 **6.** 26

7. 2 **8.** 50 **9.** 0

Study Link 10·3

1. $2 * (L + M)$, or $2 (L+M)$

2. $\frac{1}{4} * (M - (I + S))$, or $\frac{1}{4} (M - (I + S))$

3. a. Multiply N by 3 and add 5.

   b. $Q = 2N + 5$

4. a. Multiply E by 6 and add 15.

   b. $R = (E * 6) + 15$

Study Link 10·4

1. a.

Weight (lb) (w)	Cost ($) (2.50 * w)
1	2.50
3	7.50
6	15.00
10	25.00

2. a.

Gasoline (gal) (g)	Distance (mi) (24 * g)
1	24
4	96
7	168
13	312

Study Link 10·5

2. 60°F **4.** 72°F **5. a.** 70°F **b.** 67°F

Study Link 10·6

Time	Distance (yd)	
	Natasha	Derek
Start	0	10
1	6	15
2	12	20
3	18	25
4	24	30
9	54	55
10	60	60
11	66	65
12	72	70
13	80	75

Study Link 10·7

Answers vary.

Study Link 10·8

1. a. 22.0 **b.** 40.2

2. a. 85 **b.** 85

3. 21

Study Link 10·9

1. circumference **2.** area **3.** area

4. circumference **5.** 50 cm^2

6. 6 in. **7.** 5 m

8. Sample answer: The circumference is 31.4 meters, and this equals $\pi * d$, or about $3.14 * d$. Since $3.14 * 10 = 31.4$, the diameter is about 10 meters. The radius is half the diameter, or about 5 meters.

Pan-Balance Problems

SRB
228 229

In each figure below, the two pans are in perfect balance. Solve these pan-balance problems.

1. One triangle weighs

as much as _____ squares.

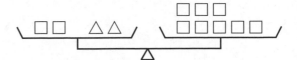

2. One cube weighs

as much as _____ marbles.

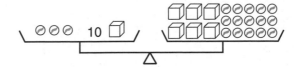

3. Two cantaloupes weigh

as much as _____ apples.

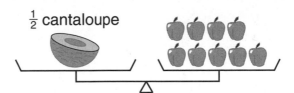

4. One X weighs

as much as _____ Ys.

| 4 X   15 Y | 6 X   7 Y |

5. One B weighs

as much as _____ Ms.

| 3 B   3 M | 1 B   9 M |

Practice

6.   4,217
       − 2,849

7.   16,000
       − 8,245

8. 11.47 − 8.896 = _____

9. 36 − 42 = _____

STUDY LINK 10·2 Pan-Balance Problems

In each figure below, the two pans are in perfect balance. Solve these pan-balance problems.

1.

One triangle weighs

as much as _____ balls.

2.

One pen weighs

as much as _____ paper clips.

3.

M	N		15 marbles

M weighs

as much as _____ marbles.

4.

5 △ □		11 □

One △ weighs

as much as _____ □s.

2 N | **20 marbles**

N weighs

as much as _____ marbles.

△ □ □ | 8 marbles

One □ weighs

as much as _____ marbles.

5.

One cup of juice weighs

as much as _____ blocks.

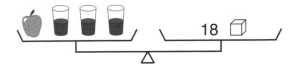 18 □

One apple weighs

as much as _____ blocks.

Practice

Fill in the missing numbers to make true sentences.

6. _____ = (7 + 45) / 2

7. ((28 / 7) + 12) / 8 = _____

8. ((14 * 3) + 14) − 6 = _____

9. _____ = (3 − 3) * ((34 / 2) * 115)

225

Name _____ Date _____ Time _____

Writing Algebraic Expressions

Complete each statement below with an algebraic expression, using the suggested variable.

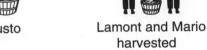

1. Lamont, Augusto, and Mario grow carrots in three garden plots. Augusto harvests two times as many carrots as the total number of carrots that Lamont and Mario harvest. So Augusto harvests

   Augusto

   Lamont and Mario harvested
   L + *M* carrots.

   _____ carrots.

2. Rhasheema and Alexis have a lemonade stand at their school fair. They promise to donate one-fourth of the remaining money (*m*) after they repay the school for lemons (*l*) and sugar (*s*). So the girls donate

   _____ dollars.

3. **a.** State in words the rule for the "What's My Rule?" table at the right.

   _____

   _____

   b. Circle the number sentence that describes the rule.

   $Q = (3 + N) * 5$ $Q = 3 * (N + 5)$ $Q = 3N + 5$

N	Q
2	11
4	17
6	23
8	29
10	35

4. **a.** State in words the rule for the "What's My Rule?" table at the right.

   _____

   _____

   b. Circle the number sentence that describes the rule.

   $R = E * 6 * 15$ $R = (E * 6) + 15$ $R = E * 15 + 6$

E	R
7	57
10	75
31	201
3	33
108	663

Practice

5. $384 * 1.5 =$ _____

6. $50.3 * 89 =$ _____

7. $\frac{843}{7} =$ _____

8. $70.4 / 8 =$ _____

227

STUDY LINK 10·4 | **Representing Rates**

Complete each table below. Then graph the data and connect the points.

1. **a.** Cherry tomatoes cost $2.50 per pound.
Rule: Cost = $2.50 * number of pounds

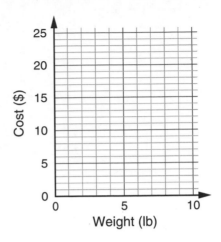

Weight (lb) (w)	Cost ($) (2.50 * w)
1	
3	
	15.00
10	

b. Plot a point to show the cost of 8 pounds.
How much would 8 pounds of cherry tomatoes cost? _____

c. Would you use the graph, the rule, or the table to find out how much
50 pounds of cherry tomatoes would cost? Explain.

_____

_____

2. **a.** Chantel is planning a trip to drive across country.
Her car uses 1 gallon of gasoline every 24 miles.
Rule: distance = 24 * number of gallons

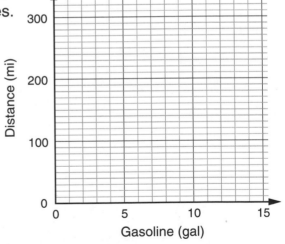

Gasoline (gal) (g)	Distance (mi) (24 * g)
1	
4	
	168
13	

b. Plot a point to show the distance the car would travel
on 6 gallons of gasoline. How many miles would it go? _____

c. Would you use the graph, the rule, or the table to find out
how far the car would travel on 9 gallons of gasoline? Explain. _____

_____

229

STUDY LINK 10·5 | **Cricket Formulas**

In 1897, the physicist, A. E. Dolbear, published an article titled "The Cricket as a Thermometer." In it he claimed that outside temperatures can be estimated by counting the number of chirps made by crickets and then by using that number in the following formula:

Outside temperature (°F) = $\frac{\text{(number of cricket chirps per minute} - 40)}{4} + 50$

1. Write a number model for the formula. _____

2. According to this formula, what is the estimated outside temperature if you count 80 chirps in a minute? _____

   Other cricket formulas exist. The following formula is supposed to work particularly well with field crickets:

   Outside temperature (°F) = (number of chirps in 15 seconds) + 37

3. Write a number model for the formula. _____

4. According to this formula, what is the estimated outside temperature if you counted 35 chirps in 15 seconds? _____

5. Compare the two formulas. If you count 30 chirps in 15 seconds, what is the estimated outside temperature for each formula?

   a. First formula: _____

   b. Second formula: _____

Practice

6. $7 - 2\frac{2}{5} =$ _____

7. $1\frac{1}{2} + 2\frac{2}{3} + 3\frac{3}{4} + \frac{1}{12} =$ _____

8. $\left(\frac{2}{3} * \frac{2}{3}\right) - \frac{2}{9} =$ _____

9. $\frac{12}{9} \div \frac{1}{3} =$ _____

STUDY LINK 10·6 | Interpreting Tables and Graphs

Natasha is 12 years old and runs an average of 6 yards per second.
Derek is 8 years old and runs about 5 yards per second. Natasha challenged
Derek to an 80-yard race and told him she would win even if he had
a 10-yard head start.

1. Complete the table showing the distances Natasha and Derek are from the
   starting line after 1 second, 2 seconds, 3 seconds, and so on.

Time (sec)	Distance (yd)	
	Natasha	Derek
Start	0	10
1		
2		20
3	18	
4		
9		55
10		
11		
12		
13		

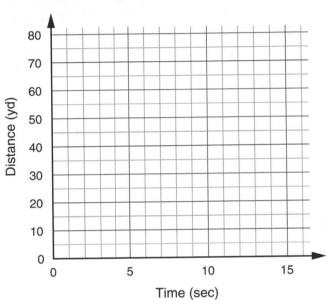

2. Use the table to write rules for the distance covered by Natasha and Derek.

   Natasha's Rule: _____

   Derek's Rule: _____

   _____

3. Graph the results of the race between Natasha and Derek on the grid above. Label each line.

4. a. Who wins the race? _____

   b. What is the winning time? _____

   c. At what time in the race did Natasha take the lead? _____

233

STUDY LINK 10·7 | **Mystery Graphs**

Create a mystery graph on the grid below. Be sure to label the horizontal and vertical axes. Describe the situation that goes with your graph on the lines provided.

_____

_____

_____

_____

_____

_____

_____

Reminder: Look for examples of ratios and bring them to school.

Name _____ Date _____ Time _____

Finding Circumferences

The formula for the circumference of a circle is:

> Circumference = π ∗ diameter, or $C = \pi * d$

SRB
187

Use the ⌜π⌝ key on your calculator to solve these problems. If your calculator doesn't have a ⌜π⌝ key, enter 3.14 each time you need π.

Find the circumference of each circle below. Show answers to the nearest tenth.

1. a.

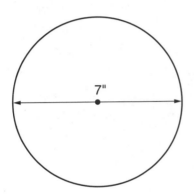

7"

Circumference ≈ _____ inches

b.

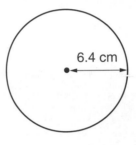

6.4 cm

Circumference ≈ _____ centimeters

2. The wheels on Will's bicycle have a diameter of about 27 inches, including the tire.

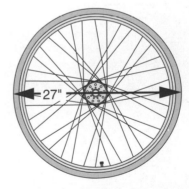

27"

a. What is the circumference of the tire?

About _____ inches

b. About how far will Will's bicycle travel if the wheels go around exactly once?

About _____ inches

3. Sofia measured the circumference of her bicycle tire. She found it was 66 inches. What is the diameter of the tire?

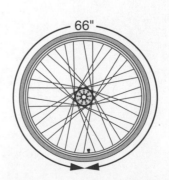

66"

About _____ inches

 STUDY LINK 10·9 | **Area and Circumference**

Circle the best measurement for each situation described below.

1. What size hat to buy (*Hint:* The hat has to fit around a head.)

      area   circumference   perimeter

2. How much frosting covers the top of a round birthday cake

      area   circumference   perimeter

3. The amount of yard that will be covered by a circular inflatable swimming pool

      area   circumference   perimeter

4. The length of a can label when you pull it off the can

      area   circumference   perimeter

Fill in the oval next to the measurement that best completes each statement.

> Area of a circle: $A = \pi * r^2$
> Circumference of a circle: $C = \pi * d$

5. The radius of a circle is about 4 cm. The area of the circle is about

   O 12 cm^2 O 39 cm^2 O 50 cm^2 O 25 cm^2

6. The area of a circle is about 28 square inches. The diameter of the circle is about

   O 3 in.   O 6 in.   O 9 in.   O 18 in.

7. The circumference of a circle is about 31.4 meters. The radius of the circle is about

   O 3 m   O 5 m   O 10 m   O 15 m

8. Explain how you found your answer for Problem 7.

_____

_____

_____

_____

_____

STUDY LINK 10·10

Unit 11: Family Letter

Volume

Unit 11 focuses on developing your child's ability to think spatially. Many times, students might feel that concepts of area and volume are of little use in their everyday lives compared with computation skills. Encourage your child to become more aware of the relevance of 2- and 3-dimensional shapes. Point out geometric solids (pyramids, cones, and cylinders) as well as 2-dimensional shapes (squares, circles, and triangles) in your surroundings.

Volume (or capacity) is the measure of the amount of space inside a 3-dimensional geometric figure. Your child will develop formulas to calculate the volume of rectangular and curved solids in cubic units. The class will also review units of capacity, such as cups, pints, quarts, and gallons. Students will use units of capacity to estimate the volume of irregular objects by measuring the amount of water each object displaces when submerged. Your child will also explore the relationship between weight and volume by calculating the weight of rice an average Thai family of four consumes in one year and by estimating how many cartons would be needed to store a year's supply.

Area is the number of units (usually squares) that can fit onto a bounded surface, without gaps or overlaps. Your child will review formulas for finding the area of rectangles, parallelograms, triangles, and circles and use these formulas in calculating the surface area of 3-dimensional shapes.

The goal of this unit is not to have students memorize formulas, but to help them develop an appreciation for using and applying formulas in various settings. By the end of this unit, your child will have had many experiences using 2- and 3-dimensional geometry.

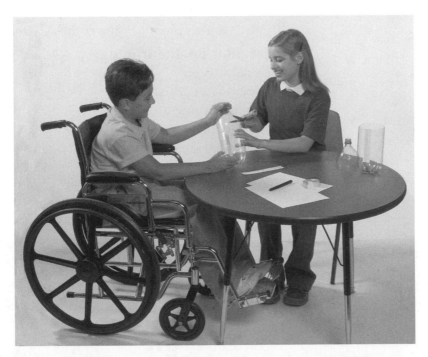

Please keep this Family Letter for reference as your child works through Unit 11.

241

Vocabulary

Important terms in Unit 11:

apex In a pyramid or cone, the vertex opposite the base.

base of a parallelogram The side of a parallelogram to which an altitude is drawn. The length of this side.

base of a prism or cylinder Either of the two parallel and congruent faces that define the shape of a prism or a cylinder.

base of a pyramid or cone The face of a pyramid or cone that is opposite its apex.

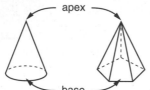

calibrate To divide or mark a measuring tool, such as a thermometer, with gradations.

cone A geometric solid with a circular base, a vertex (apex) not in the plane of the base, and all of the line segments with one endpoint at the apex and the other endpoint on the circumference of the base.

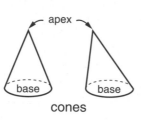

cones

cube A polyhedron with 6 square faces. A cube has 8 vertices and 12 edges.

cylinder A geometric solid with two congruent, parallel circular regions for bases, and a curved face formed by all the segments with an endpoint on each circle that are parallel to the segment with endpoints at the center of the circles.

cylinder

edge A line segment where two faces of a polyhedron meet.

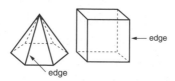

edge

face A flat surface on a polyhedron.

geometric solid The surface or surfaces that make up a 3-dimensional shape, such as a prism, pyramid, cylinder, cone, or sphere. Despite its name, a geometric solid is hollow; it does not contain the points in its interior.

polyhedron A 3-dimensional shape formed by polygons with their interiors (faces) and having no holes.

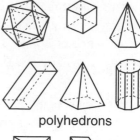

polyhedrons

prism A polyhedron with two parallel and congruent polygonal regions for bases and lateral faces formed by all the line segments with endpoints on corresponding edges of the bases. The lateral faces are all parallelograms. Prisms get their names from the shape of their bases.

triangular prism   rectangular prism

pyramid A polyhedron made up of any polygonal region for a base, a point (apex) not in the plane of the base, and all of the line segments with one endpoint at the apex and the other on an edge of the base. All the faces except perhaps the base are triangular. Pyramids get their names from the shape of their base.

square pyramid

regular polyhedron A polyhedron whose faces are all congruent regular polygons and in which the same number of faces meet at each vertex.

tetrahedron   cube   octahedron

dodecahedron   icosahedron

The five regular polyhedrons

sphere The set of all points in space that are a given distance from a given point. The given point is the center of the sphere, and the given distance is the radius.

surface area A measure of the surface of a 3-dimensional figure.

vertex (vertices or vertexes) The point where the rays of an angle, the sides of a polygon, or the edges of a polyhedron meet.

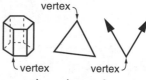

vertex   vertex   vertex

Do-Anytime Activities

To work with your child on the concepts taught in this unit and in previous units, try these interesting and rewarding activities.

1. Have your child compile a 2- and 3-dimensional shapes portfolio or create a collage of labeled shapes. Images can be taken from newspapers, magazines, photographs, and so on.

2. **Explore Kitchen Measures**
   The most common use of measuring volume is cooking. Work with your child to make a favorite recipe. (Doubling the recipe can be good practice in computing with fractions.) Ask your child to use measuring spoons and cups to find the capacity of various containers. The data can be organized in a table.

Container	Capacity
Coffee mug	$1\frac{1}{4}$ cups
Egg cup	3 tablespoons

Building Skills through Games

In Unit 11, your child will practice operations with whole numbers and geometry skills by playing the following games. Detailed instructions for each game are in the *Student Reference Book* or the journal:

Name That Number See *Student Reference Book,* page 325.
This is a game for two or three players using the Everything Math Deck or a complete deck of number cards. Playing *Name That Number* helps students review operations with whole numbers, including the order of operations.

3-D Shape Sort See *Student Reference Book,* page 332.
This game is similar to *Polygon Capture*. Partners or 2 teams each with 2 players need 16 Property cards and 12 Shape cards to play. *3-D Shape Sort* gives students practice identifying properties of 3-dimensional shapes.

Rugs and Fences See journal page 380.
This game uses 32 Polygon cards and 16 Area and Perimeter cards and is played by partners. *Rugs and Fences* gives students practice finding the area and perimeter of polygons.

As You Help Your Child with Homework

As your child brings assignments home, you might want to go over the instructions together, clarifying them as necessary. The answers listed below will guide you through some of this unit's Study Links.

Study Link 11·1

1. Answers vary.

2. D

Study Link 11·2

1.

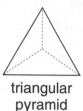

triangular
pyramid

base

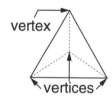

vertex

vertices

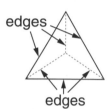

edges

edges

Study Link 11·3

Sample answers:

1. 2.8 cm; 4.3 cm; 24.6 cm²; 105.9 cm³

3a. 30 * 30 * 18 = 16,200

5. more; 283,500,000 cm³

7. $5\frac{3}{8}$

Study Link 11·4

1. <   **2.** <   **3.** >

4. Because both pyramids have the same height, compare the areas of the bases. The base of the square pyramid has an area of 5 * 5 = 25 m². The base area of the triangular pyramid is $\frac{1}{2}$ * 5 * 5 or $12\frac{1}{2}$ m².

5. $10\frac{16}{27}$ **6.** $1\frac{11}{21}$ **7.** 600,000   **8.** 25.39

Study Link 11·5

Most of the space taken up by a handful of cotton is air between the fibers.

Study Link 11·6

1. >   **2.** =   **3.** <

4. <   **5.** <   **6.** =

7. cubic inches

8. gallons

9. gallons

10. milliliters

11. cubic centimeters

12. capacity

13. volume

14. −250   **15.** 137,685

16. $10\frac{2}{5}$ **17.** 0.48

Study Link 11·7

1. 88 in²; Sample answer: I found the area of each of the 6 sides and then added them together.

2. Yes. A 4 in. by 4 in. by $3\frac{1}{2}$ in. box has a volume of 56 in³ and a surface area of 88 in².

3. Volume: 502.4 cm³; Surface area: 351.7 cm²

4. Volume: 216 in³; Surface area: 216 in²

STUDY LINK
11·1

Cube Patterns

There are four patterns below. Three of the patterns can be folded to form a cube.

1. Guess which one of the patterns below cannot be folded into a cube.

My guess: Pattern _____ (A, B, C, or D) cannot be folded into a cube.

2. Cut on the solid lines, and fold the pattern on the dashed lines to check your guess. Did you make the correct guess? If not, try other patterns until you find the one that does not form a cube.

My answer: Pattern _____ (A, B, C, or D) cannot be folded into a cube.

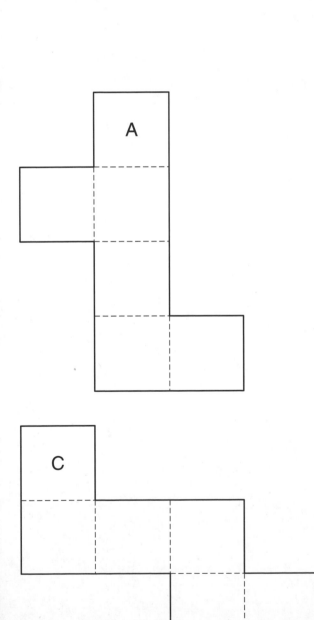

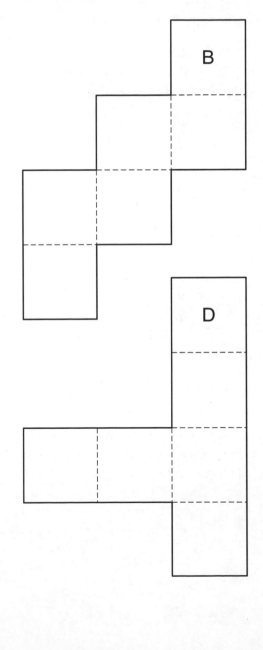

STUDY LINK
11·2

Comparing Geometric Solids

Name the figures, and label their bases, vertices, and edges.

Geometric Solids			
Name	**Bases**	**Vertices**	**Edges**
Example cube	base base	vertices vertices	edges edges edges edges
1. _____			
2. _____			
3. _____			

247

STUDY LINK
11·3

Volume of Cylinders

SRB
194
197 198

Use these two formulas to solve the problems below.

Formula for the Volume of a Cylinder	Formula for the Area of a Circle
$V = B * h$	$A = \pi * r^2$
where V is the volume of the cylinder, B is the area of the cylinder's base, and h is the height of the cylinder.	where A is the area of the circle and r is the length of the radius of the circle.

1. Find the smallest cylinder in your home. Record its dimensions, and calculate its volume.

   radius = _____   height = _____

   Area of base = _____   Volume = _____

2. Find the largest cylinder in your home. Record its dimensions, and calculate its volume.

   radius = _____   height = _____

   Area of base = _____   Volume = _____

3. Write a number model to estimate the volume of:

   a. Your toaster _____

   b. Your television _____

4. Is the volume of the largest cylinder more
   or less than the volume of your toaster?   _____

   About how much more or less?   _____

5. Is the volume of the largest cylinder more or
   less than the volume of your television set?   _____

   About how much more or less?   _____

Practice

6. $6\frac{1}{3} * \frac{2}{5} =$ _____   7. $10\frac{6}{8} * \frac{1}{2} =$ _____   8. $4 - 2.685 =$ _____

249

A Displacement Experiment

Try this experiment at home.

Materials
- ☐ drinking glass
- ☐ water
- ☐ 2 large handfuls of cotton
  (Be sure to use real cotton. Synthetic materials will not work.)

Directions

◆ Fill the drinking glass almost to the top with water.

◆ Put the cotton, bit by bit, into the glass. Fluff it as you go.

If you are careful, you should be able to fit all of the cotton into the glass without spilling a drop of water.

Think about what you know about displacement and volume. Why do you think you were able to fit the cotton into the glass without the water overflowing?

_____

_____

_____

_____

_____

_____

_____

_____

_____

_____

_____

STUDY LINK
11·6

Units of Volume and Capacity

Write >, <, or = to compare the measurements below.

1. 5 cups _____ 1 quart   **2.** 30 mL _____ 30 cm³   **3.** 1 quart _____ 1 liter

4. 15 pints _____ 8 quarts   **5.** 100 cm³ _____ 1 gallon   **6.** 10 cups _____ 5 pints

Circle the unit you would use to measure each of the following.

7. The volume of a square pyramid

  gallons   cubic inches   ounces   meters

8. The amount of milk a fifth grader drinks in a week

  gallons   milliliters   ounces   meters

9. The amount of water used to fill a swimming pool

  gallons   milliliters   ounces   meters

10. The amount of penicillin given in a shot

  gallons   milliliters   liters   meters

11. The volume of a rectangular prism

  gallons   cubic centimeters   liters   meters

12. Would you think of volume or capacity if
you wanted to know how much juice a jug holds? _____

13. Would you think of volume or capacity if you wanted to
know how much closet space a stack of boxes would take up? _____

Practice

14. $-200 + (-50) =$ _____   **15.** $685 * 201 =$ _____

16. $13\frac{1}{5} - 2\frac{4}{5} =$ _____   **17.** $3.84 \div 8 =$ _____

STUDY LINK 11·7

Volume and Surface Area

Area of rectangle:
$A = l * w$

Volume of rectangular prism:
$V = l * w * h$

Circumference of circle:
$c = \pi * d$

Area of circle:
$A = \pi * r^2$

Volume of cylinder:
$V = \pi * r^2 * h$

1. Kesia wants to give her best friend a box of chocolates. Figure out the least number of square inches of wrapping paper Kesia needs to wrap the box. (To simplify the problem, assume that she will cover the box completely with no overlaps.)

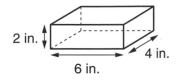

2 in.
4 in.
6 in.

Amount of paper needed: _____

Explain how you found the answer.

_____

_____

_____

2. Could Kesia use the same amount of wrapping paper to cover a box with a larger volume than the box in Problem 1? _____ Explain.

_____

_____

Find the volume and the surface area of the two figures in Problems 3 and 4.

3. Volume:

_____

Surface area:

_____

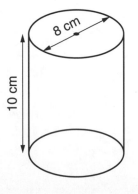

8 cm

10 cm

4. Volume:

_____

Surface area:

_____

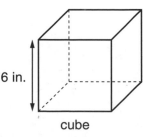

6 in.

cube

257

STUDY LINK 11·8 Unit 12: Family Letter

Probability, Ratios, and Rates

A **ratio** is a comparison of two quantities with the same unit. For example, if one house has a floor area of 2,000 ft², and a second house has a floor area of 3,000 ft², the ratio of the areas is 2,000 to 3,000, or 2 to 3, simplified.

To prepare students for working with ratios in algebra, the class will review the meanings and forms of ratios and will solve number stories involving ratios of part of a set to the whole set. Your child will find, write, and solve many number models (equations) for ratio problems.

Your child will continue to use the American Tour section of the *Student Reference Book* as part of the discussion of ratios. We will also be doing projects based on information in the American Tour.

A **rate** is a comparison of two quantities with different units. For example, speed is expressed in miles per hour. In our study of rates, students will determine their own heart rates (heartbeats per minute). Then they will observe the effect of exercise on heart rate and represent the class results graphically.

We will continue our study of probability by looking at situations in which a sequence of choices is made. For example, if a menu offers you 2 choices of appetizer, 4 choices of entrée, and 3 choices of dessert, and you choose one of each kind, there are 2 * 4 * 3 or 24 different possible combinations for your meal. If all the choices were equally appealing (which is unlikely), and you chose at random, the probability of any one combination would be $\frac{1}{24}$.

Your child will play *Frac-Tac-Toe,* which was introduced in Unit 5, as well as a new game, *Spoon Scramble,* to practice operations and equivalencies with fractions, decimals, and percents.

You can help your child by asking questions about homework problems; by pointing out fractions, percents, and ratios that you encounter in everyday life; and by playing *Frac-Tac-Toe* and *Spoon Scramble* to sharpen his or her skills.

Please keep this Family Letter for reference as your child works through Unit 12.

Vocabulary

Important terms in Unit 12:

common factor Any number that is a factor of two or more counting numbers. The common factors of 18 and 24 are 1, 2, 3, and 6.

equally likely outcomes Outcomes of a chance experiment or situation that have the same probability of happening. If all the possible outcomes are equally likely, then the probability of an event is equal to: $\frac{\text{number of favorable outcomes}}{\text{number of possible outcomes}}$

factor tree A method used to obtain the prime factorization of a number. The original number is written as a product of factors. Then each of these factors is written as a product of factors, and so on, until the factors are all prime numbers. A factor tree looks like an upside down tree with the root (the original number) at the top, and the leaves (the factors) beneath it.

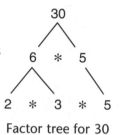

Factor tree for 30

greatest common factor The largest factor that two or more counting numbers have in common. For example, the common factors of 24 and 36 are 1, 2, 3, 4, 6, and 12. Thus, the greatest common factor of 24 and 36 is 12.

least common multiple The smallest number that is a multiple of two or more numbers. For example, while some common multiples of 6 and 8 are 24, 48, and 72, the least common multiple of 6 and 8 is 24.

multiplication counting principle A way of determining the total number of possible outcomes for two or more separate choices. Suppose, for example, you roll a die and then flip a coin. There are 6 choices for which number on the die lands up and 2 choices for which side of the coin shows. Then there are 6 * 2, or 12 possible outcomes all together: (1,H), (1,T), (2,H), (2,T), (3,H), (3,T), (4,H), (4,T), (5,H), (5,T), (6,H), (6,T).

prime factorization A counting number expressed as a product of prime number factors. For example, the prime factorization of 24 is 2 * 2 * 2 * 3, or $2^3 * 3$.

probability A number from 0 to 1 that tells the chance that an event will happen. For example, the probability that a fair coin will show heads is $\frac{1}{2}$. The closer a probability is to 1, the more likely it is that the event will happen. The closer a probability is to 0, the less likely it is that the event will happen.

rate A comparison by division of two quantities with unlike units. For example, traveling 100 miles in 2 hours is an average rate of 100 mi/2 hr, or 50 miles per hour. In this case, the rate compares distance (miles) to time (hours).

ratio A comparison by division of two quantities with the same units. Ratios can be fractions, decimals, percents, or stated in words. Ratios can also be written with a colon between the two numbers being compared. For example, if a team wins 3 out of 5 games played, the ratio of wins to total games can be written as $\frac{3}{5}$, 3/5, 0.6, 60%, 3 to 5, or 3:5 (read "three to five").

tree diagram A network of points connected by line segments and containing no closed loops. Factor trees are tree diagrams used to factor numbers. Probability trees are tree diagrams used to represent probability situations in which there is a series of events.

The first tree diagram below represents flipping one coin two times. The second tree diagram below shows the prime factorization of 30.

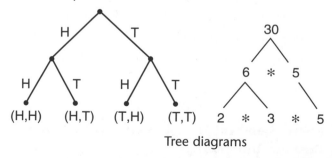

Tree diagrams

Do-Anytime Activities

To work with your child on the concepts taught in this unit and in previous units, try these interesting and rewarding activities:

1. Identify different ratios, and ask your child to write each ratio using words, a fraction, a decimal, a percent, and a colon. For example, the ratio of 1 adult for every 5 students could be written as 1 to 5, $\frac{1}{5}$, 0.2, 20%, or 1:5.

2. Play one of the games in this unit with your child: *Frac-Tac-Toe, Name That Number,* or *Spoon Scramble.*

3. Read the book *Jumanji* with your child, and review the possible outcomes when rolling two dice. Ask your child to verify the probabilities of rolling certain number combinations by recording the outcomes for 100 rolls of a pair of dice.

4. Identify rate situations in everyday life, and ask your child to solve problems involving rates. For example, find the number of miles your car travels for each gallon of gas, or find the number of calories that are burned each hour or minute for different types of sports activities.

Building Skills through Games

In Unit 12, your child will practice skills with probability, ratios, and rates by playing the following games. For detailed instructions, see the *Student Reference Book.*

Frac-Tac-Toe See *Student Reference Book,* pages 309–311. This is a game for two players. Game materials include 4 each of the number cards 0–10, pennies or counters of two colors, a calculator, and a gameboard. The gameboard is a 5-by-5 number grid that resembles a bingo card. Several versions of the gameboard are shown in the *Student Reference Book. Frac-Tac-Toe* provides students with practice in converting fractions to decimals and percents.

Name That Number See *Student Reference Book,* page 325. This is a game for two or three players. Game materials include the Everything Math Deck or a complete deck of number cards. Playing *Name That Number* provides students with practice in working with operations and in using the order of operations.

Spoon Scramble See *Student Reference Book,* page 330. This is a game for four players using 3 spoons and a deck of 16 *Spoon Scramble* Cards. *Spoon Scramble* provides students with practice identifying equivalent expressions for finding a fraction, a decimal, or a percent of a number.

As You Help Your Child with Homework

As your child brings assignments home, you might want to go over the instructions together, clarifying them as necessary. The answers listed below will guide you through this unit's Study Links.

Study Link 12·1

1. a. 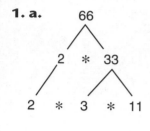 **b.**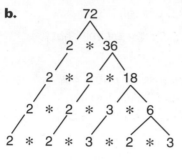

2. a. $\frac{10}{33}$ **b.** $\frac{11}{12}$ **c.** $\frac{5}{18}$

3. $250 = 5 * 5 * 5 * 2$

4. a. 32   **b.** 49   **5.** $\frac{2}{3}$

Study Link 12·2

1. $5 * 5 = 25$

2.

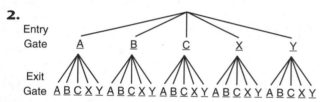

Entry Gate   A   B   C   X   Y

Exit Gate   A B C X Y A B C X Y A B C X Y A B C X Y A B C X Y

3. No; Sample answer: Some gates will probably be used more than other gates.

4. 20

5. a.

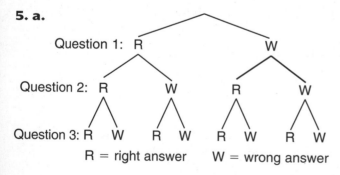

Question 1: R   W

Question 2: R   W   R   W

Question 3: R   W   R   W   R   W   R   W

R = right answer   W = wrong answer

b. $\frac{1}{8}$

Study Link 12·3

1. Sixteen out of twenty-five

2. $\frac{16}{25}$ **3.** 64%   **4.** 16:25

5. 23:50; 0.46 of the cars were blue

6. $\frac{2}{3}$; 6:9; $66\frac{2}{3}$% of the people were swimmers

7. 7 out of 8; 35:40 of the caps sold were baseball caps

Study Link 12·4

1. a. 4   **b.** 16   **2.** 15

3. 16   **4.** 8   **5.** 32

6. 98 R38   **7.** 9,016   **8.** 90.54

Study Link 12·5

1. 8   **2.** 24   **3.** 45

4. 60   **5.** 20   **6.** 26

7. $\frac{2}{5} = \frac{\square}{115}$; 46 students

9. $\frac{1.50}{3} = \frac{\square}{90}$; $45.00

11. 216   **12.** 729

Study Link 12·6

1. a.

Number of Spiders	27,000	54,000	81,000	108,000	135,000
Pounds of Spider Web	1	2	3	4	5

b. 270,000

3. 1,000   **4.** 930   **5.** $7\frac{1}{2}$, or 7.5

Study Link 12·7

1. $3\frac{3}{4}$ in.   **3.** $1\frac{3}{4}$ lb   **5.** $20\frac{7}{8}$ in.

7. $50\frac{2}{5}$ kg   **9.** 34   **11.** 180

Study Link 12·8

2. 8 lunches

4. a. 1 to 1   **b.** 26 to 104, or $\frac{1}{4}$ **c.** 8 to 16, or $\frac{1}{2}$

5. $3\frac{4}{7}$ **6.** 5   **7.** 12.5   **8.** 8

STUDY LINK 12·1 | **Factor Trees**

1. Make factor trees and find the prime factorization for the following numbers.

   Example: 20

   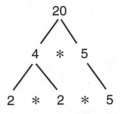

   20 = 2 * 2 * 5

   a. 66   **b.** 72

   66 = _____   72 = _____

2. Write each fraction in simplest form. Use factor trees to help you. Show your work.

   a. $\frac{20}{66}$ = _____   **b.** $\frac{66}{72}$ = _____   **c.** $\frac{20}{72}$ = _____

3. Find the prime factorization for 250. _____

4. **a.** Circle the number that has the most prime factors.

   63   32   49   100

   b. Which has the fewest prime factors? _____

5. Simplify the fraction to the right. $\frac{150}{225}$ = _____

Practice

6. $\frac{1}{4}$ * 36 = _____   7. 0.25 * 360 = _____

8. $\frac{1}{3}$ * 90 = _____   9. $33\frac{1}{3}$% of 90 = _____

STUDY LINK 12·3 | Ratios

Ratios can be stated or written in a variety of ways. Sometimes a ratio is easier to understand or will make more sense if it is rewritten in another form.

Example: In a group of 25 students, 16 students walk to school and 9 take a bus. The ratio of students who take a bus compared to all students in the group can be expressed in the following ways:

◆ With words: Nine out of twenty-five students take a bus.

◆ With a fraction: $\frac{9}{25}$ of the students take a bus.

◆ With a percent: 36% of the students take a bus.

◆ With a colon between the two numbers being compared: The ratio of students who take a bus to all students in the group is 9:25 (nine out of twenty-five).

Revise the above statements to express the ratio of students who walk to school to all students.

1. With words: _____ students walk to school.

2. With a fraction: _____ of the students walk to school.

3. With a percent: _____ of the students walk to school.

4. With a colon: The ratio of students
who walk to school to all students is _____ .

In each problem, fill in the ovals next to each correct ratio.

5. Fifty cars drove past in 10 minutes. Twenty-three cars were blue.

  O 23:50 of the cars were blue.   O 23% of the cars were blue.   O 0.46 of the cars were blue.

6. In a group of 9 people, 6 were swimmers.

  O $\frac{2}{3}$ of the people were swimmers.   O 6:9 of the people were swimmers.   O $66\frac{2}{3}$% of the people were swimmers.

7. In a sports shop, 35 of the 40 caps sold the day before the World Series were baseball caps.

  O 7 out of 8 caps sold were baseball caps.   O 35% of the caps sold were baseball caps.   O 35:40 of the caps sold were baseball caps.

267

STUDY LINK
12·4 **Ratio Problems**

1. Draw 20 tiles so that 2 out of 10 tiles are white and the rest are shaded.

   a. How many tiles are white? _____ tiles

   b. How many tiles are shaded? _____ tiles

2. Draw 9 shaded tiles.

   Add white tiles until 2 out of 5 tiles are white.

   How many tiles are there in all? _____ tiles

3. Imagine 48 tiles. If 4 out of 12 tiles
   are white, how many tiles are white? _____ tiles

4. There are 24 players on the soccer team. Two out of
   every 3 players have not scored a goal yet this year.
   How many players have scored a goal this year? _____ players

5. For every 8 spelling tests Justine took, she earned
   3 perfect scores. If Justine earned 12 perfect
   scores this year, how many spelling tests did she take? _____ tests

---

Practice

6. $92\overline{)9,054}$ → _____

7. $98 * 92 =$ _____

8. $90.16 + 0.38 =$ _____

9. $90.54 * 10^2 =$ _____

STUDY LINK
12·5

Ratio Problems

SRB
106–109
243–245

Find the missing number.

1. $\frac{1}{5} = \frac{x}{40}$ $x =$ _____

2. $\frac{2}{3} = \frac{16}{y}$ $y =$ _____

3. $\frac{5}{6} = \frac{m}{54}$ $m =$ _____

4. $\frac{1}{4} = \frac{15}{n}$ $n =$ _____

5. $\frac{5}{8} = \frac{f}{32}$ $f =$ _____

6. $\frac{13}{50} = \frac{g}{100}$ $g =$ _____

Write a number model for each problem. Then solve the problem.

7. Of the 115 students in the sixth grade, 2 out of 5 belong to the Drama Club.
How many students are members of the Drama Club?

Number model: _____   Answer: _____
(unit)

8. Three out of 4 students at Highland School ordered a hot lunch today. There
are 156 students at the school. How many students ordered a hot lunch?

Number model: _____   Answer: _____
(unit)

9. Gina and the other members of her troop sell cookies for $3 a box. For each
box they sell, the troop earns $1.50. One week, Gina's troop sold $90 worth
of cookies. How much did the troop earn?

Number model: _____   Answer: _____

10. 30% of the tickets sold by a movie theater for the Friday night show were
children's tickets at $4 each. The rest of the tickets were sold at the full price
of $8.50. The movie theater collected $360 just for the children's tickets. How many
tickets did they sell in all?

Number model: _____   Answer: _____
(unit)

Practice

11. $6^3 =$ _____

12. $3^6 =$ _____

13. $6^3 * 10^2 =$ _____

271

STUDY LINK 12·6 | Rates

SRB
103–105

Complete each table using the given information. Then answer the question below each table.

1. It would take 27,000 spiders, each spinning a single web, to produce a pound of spider webs.

a.

Number of Spiders	27,000	54,000			
Pounds of Spider Webs	1	2	3	4	5

b. At this rate, how many spiders, each spinning a single web, would be needed to produce 10 pounds of spider webs? _____ spiders

2. It used to be thought that the deer botfly flies so fast that it is almost invisible to the human eye. It has since been tested, and scientists found that it actually flies about 25 miles per hour.

a.

Miles	25				
Hours	1	2	3	4	5

b. At this rate, about how far could a deer botfly travel in 1 minute? _____ mile(s)

Solve the following rate problems. Make a table if it will help you.

3. About 50 gallons of maple sap are needed to make 1 gallon of maple syrup. How many gallons of maple sap are needed to make 20 gallons of maple syrup?

About _____ gallons

4. For 186 days a year, the sun is not visible at the North Pole. During a 5-year period, about how many days is the sun not visible?

About _____ days

5. In a beehive, about $1\frac{1}{2}$ ounces of beeswax are used to build a honeycomb that holds 4 pounds of honey. How much beeswax is needed to build a honeycomb that could hold 20 pounds of honey?

About _____ ounces

Source: *2201 Fascinating Facts*

273

STUDY LINK 12·7

Operations with Fractions

1. In the Malagasay Indian tribes, it is against the law for a son to be taller than his father. If a son is taller, he must give his father money or an ox. Suppose a father is 5 feet $10\frac{1}{2}$ inches tall and his son is 5 feet $6\frac{3}{4}$ inches tall. How many more inches can the son grow before he is as tall as his father?

_____
(unit)

2. In the state of Indiana, it is illegal to travel on a bus within 4 hours of eating garlic. If you lived in Indiana and had eaten a bowl of pasta with garlic bread $2\frac{1}{3}$ hours ago, how many more hours would you need to wait before you could legally travel on a bus?

_____
(unit)

3. In Idaho, it is against the law to give a person a box of candy that weighs more than 50 pounds. It is Valentine's Day, and you give your mother a box of candy that weighs $48\frac{1}{4}$ pounds. How much more could the box weigh without breaking the law?

_____
(unit)

4. The body of an average jellyfish is about $\frac{9}{10}$ water. What fraction of the jellyfish is not water?

_____

5. The world record for a jump by a frog is 19 feet $3\frac{1}{8}$ inches. How much farther would a frog need to jump to set a new world record of 7 yards?

_____
(unit)

6. The maximum length for a typical king cobra is about $5\frac{4}{5}$ meters. If 6 of these snakes were lined up end to end, how far would they stretch?

_____
(unit)

7. An average trumpeter swan weighs about $16\frac{4}{5}$ kilograms. What is the approximate weight of 3 average trumpeter swans?

_____
(unit)

Sources: *The Top 10 of Everything; Beyond Belief!*

Practice

8. $(4 * 4) + \frac{4}{4} =$ _____

9. $4! + 4 + 4 + \sqrt{4} =$ _____

10. 75% of 12 = _____

11. 50% of 360 = _____

275

STUDY LINK
12·8

Rate and Ratio Problems

SRB
104 105

1. The average American eats about 250 eggs per year. At this rate, about how many eggs will the average American eat in . . .

   a. five years? _____
   (unit)

   b. $\frac{1}{12}$ of a year? _____
   (unit)

2. The average fifth grader can eat $\frac{3}{8}$ of a pizza for lunch. At this rate, how many lunches will it take for an average fifth grader to eat the equivalent of 3 whole pizzas? _____
   (unit)

3. In 1975, a man in Washington state ate 424 clams in 8 minutes. At this rate, how many would he eat . . .

   a. in $\frac{1}{4}$ of this time? _____
   (unit)

   b. in $2\frac{1}{2}$ times as much time? _____
   (unit)

4. A deck has 52 playing cards. In two decks,

   a. what is the ratio of 2s to 10s? _____

   b. what is the ratio of Hearts to the total number of playing cards? _____

   c. what is the ratio of Jacks to Kings and Queens? _____

Practice

5. $3\frac{4}{7} * \frac{8}{8} =$ _____

6. $3n + 2n = 25$

   $n =$ _____

7. $25 = 2n$

   $n =$ _____

8. $12.5 * n = 100$

   $n =$ _____

STUDY LINK
12·9

End-of-Year Family Letter

Congratulations!

By completing *Fifth Grade Everyday Mathematics,* your child has accomplished a great deal. Thank you for your support!

This Family Letter provides a resource throughout your child's vacation. It includes an extended list of Do-Anytime Activities, directions for games that can be played at home, a list of mathematics-related books to check out over vacation, and a preview of what your child will be learning in *Sixth Grade Everyday Mathematics.* Enjoy your vacation!

Do-Anytime Activities

Mathematics means more when it is rooted in real-life situations. To help your child review many of the concepts he or she has learned in fifth grade, we suggest the following activities for you to do together over vacation. These activities will help your child build on the skills he or she has learned this year and will help prepare him or her for *Sixth Grade Everyday Mathematics.*

1. Review multiplication facts. For example, include basic facts such as 7 * 8 = 56, and extended facts, such as 70 * 8 = 560 and 70 * 80 = 5,600.

2. Create opportunities to work with rulers, yardsticks, metersticks, tape measures, and scales. Have your child measure items using metric and U.S. customary units.

3. Ask your child to solve multiplication and division problems that are based on real-life situations. Vary the problems so that some are suitable for mental computation, some require paper-and-pencil calculation, and some require the use of a calculator.

4. Practice using percents by asking your child to calculate sales tax, percent discounts, sports statistics, and so on.

5. Continue the American Tour by reading about important people, events, inventions, explorations, and other topics in American history. Focus on data displays such as bar, line, and circle graphs, and on color-coded maps.

Building Skills through Games

The following section lists rules for games that can be played at home. The number cards used in some games can be made from 3" by 5" index cards.

Factor Captor

1. To start the first round, Player 1 (James) chooses a 2-digit number on the number grid. James covers it with a counter and records the number on scratch paper. This is James's score for the round.

2. Player 2 (Emma) covers all the factors of James's number. Emma finds the sum of the factors, and records it on scratch paper. This is Emma's score for the round.

   A factor may only be covered once during a round.

3. If Emma missed any factors, James can cover them with counters and add them to his score.

4. In the next round, players switch roles. Player 2 (Emma) chooses a number that is not covered by a counter. Player 1 (James) covers all factors of that number.

5. Any number that is covered by a counter is no longer available and may not be used again.

6. The first player in a round may not cover a number less than 10, unless no other numbers are available.

7. Play continues with players trading roles in each round, until all numbers on the grid have been covered. Players then use their calculators to find their total scores. The player with the higher total score wins the game.

2-4-5-10 Frac-Tac-Toe

Advance Preparation: Separate the cards into two piles—a numerator pile and a denominator pile. Place two each of the 2, 4, 5, and 10 cards in the denominator pile. All other cards are placed on the numerator pile.

Shuffle the cards in each pile. Place the piles facedown. When the numerator pile is completely used, reshuffle that pile, and place it facedown. When the denominator pile is completely used, turn it over, and place it facedown without reshuffling it.

1. Players take turns. When it is your turn:

   ◆ Turn over the top card from each pile to form a fraction (numerator card over denominator card).

   ◆ Try to match the fraction shown with one of the grid squares on the gameboard. (Use either of the gameboards shown.) If a match is found, cover that grid square with your counter and your turn is over. If no match is found, your turn is over.

Factor Captor
Grid 1

1	2	2	2	2	2
2	3	3	3	3	3
3	4	4	4	4	5
5	5	5	6	6	7
7	8	8	9	9	10
10	11	12	13	14	15
16	18	20	21	22	24
25	26	27	28	30	32

2-4-5-10 Frac-Tac-Toe
Gameboards

> 1.0	0 or 1	> 2.0	0 or 1	> 1.0
0.1	0.2	0.25	0.3	0.4
> 1.5	0.5	> 1.5	0.5	> 1.5
0.6	0.7	0.75	0.8	0.9
> 1.0	0 or 1	> 2.0	0 or 1	> 1.0

>100%	0% or 100%	>200%	0% or 100%	>100%
10%	20%	25%	30%	40%
>100%	50%	>200%	50%	>100%
60%	70%	75%	80%	90%
>100%	0% or 100%	>200%	0% or 100%	>100%

2. To change the fraction shown by the cards to a decimal or percent, players *may* use a calculator.

3. **Scoring** The first player covering three squares in a row in any direction (horizontal, vertical, diagonal) is the winner.

Variations:

◆ For a *2-4-8* game, place two each of the 2, 4, and 8 cards in the denominator pile. Use the gameboards shown in the margin.

◆ For a *3-6-9* game, place two each of the 3, 6, and 9 cards in the denominator pile. Use the gameboards shown in the margin.

Multiplication Bull's-eye

1. Shuffle a deck of number cards (4 each of the numbers 0–9) and place them facedown on the playing surface.

2. Players take turns. When it is your turn:

◆ Roll a six-sided die. Look up the target range of the product in the table.

◆ Take four cards from the top of the deck.

◆ Use the cards to try and form two numbers whose product falls within the target range. **Do not use a calculator.**

◆ Multiply the two numbers on your calculator to determine whether the product falls within the target range. If it does, you have hit the bull's-eye and score 1 point. If it doesn't, you score 0 points.

◆ Sometimes it is impossible to form two numbers whose product falls within the target range. If this happens, you score 0 points for that turn.

3. The game ends when each player has had five turns.

4. The player scoring more points wins the game.

Example:

Tom rolls a 3, so the target range of the product is from 1,001 to 3,000.

He turns over a 5, a 7, a 2, and a 9.

Tom uses estimation to try to form two numbers whose product falls within the target range, for example, 97 and 25.

He finds the product on the calculator: $97 * 25 = 2,425$.

Because the product is between 1,001 and 3,000, Tom has hit the bull's-eye and scores 1 point.

Some other possible winning products from the 5, 7, 2, and 9 cards are: $25 * 79$, $27 * 59$, $9 * 257$, and $2 * 579$.

Number on Die	Target Range of Product
1	500 or less
2	501–1,000
3	1,001–3,000
4	3,001–5,000
5	5,001–7,000
6	more than 7,000

2-4-8 Frac-Tac-Toe Gameboards

>2.0	0 or 1	>1.5	0 or 1	>2.0
1.5	0.125	0.25	0.375	1.5
>1.0	0.5	0.25 or 0.75	0.5	>1.0
2.0	0.625	0.75	0.875	2.0
>2.0	0 or 1	1.125	0 or 1	>2.0

>200%	0% or 100%	>150%	0% or 100%	>200%
150%	$12\frac{1}{2}$%	25%	$37\frac{1}{2}$%	150%
>100%	50%	25% or 75%	50%	>100%
200%	$62\frac{1}{2}$%	75%	$87\frac{1}{2}$%	200%
>200%	0% or 100%	$112\frac{1}{2}$%	0% or 100%	>200%

3-6-9 Frac-Tac-Toe Gameboards

>1.0	0 or 1	$0.\overline{1}$	0 or 1	>1.0
$0.1\overline{6}$	$0.\overline{2}$	$0.\overline{3}$	$0.\overline{3}$	$0\overline{4}$
>2.0	$0.\overline{5}$	>1.0	$0.\overline{6}$	>2.0
$0.\overline{6}$	$0.\overline{7}$	$0.8\overline{3}$	$0.\overline{8}$	$1.\overline{3}$
>1.0	0 or 1	$1.\overline{6}$	0 or 1	>1.0

>100%	0% or 100%	11.1%	0% or 100%	>100%
$16\frac{2}{3}$%	22.2%	$33\frac{1}{3}$%	33.3%	44.4%
>200%	55.5%	>100%	66.6%	>200%
$66\frac{2}{3}$%	77.7%	$83\frac{1}{3}$%	88.8%	$133\frac{1}{3}$%
>100%	0% or 100%	$166\frac{2}{3}$%	0% or 100%	>100%

Vacation Reading with a Mathematical Twist

Books can contribute to children's learning by presenting mathematics in a combination of real-world and imaginary contexts. The titles listed below were recommended by teachers who use *Everyday Mathematics* in their classrooms. They are organized by mathematical topics. Visit your local library and check out these mathematics-related books with your child.

Numeration

The Rajah's Rice: A Mathematical Folktale from India by David Barry

Operations and Computation

Counting on Frank by Rod Clement

Data and Chance

Jumanji by Chris Van Allsburg

Geometry

A Cloak for the Dreamer by Aileen Friedman; *Flatland* by Edwin Abbott; *The Boy Who Reversed Himself* by William Sleator

Measurement and Reference Frames

Spaghetti and Meatballs for All!: A Mathematical Story by Marilyn Burns; *Mr. Archimedes' Bath* by Pamela Allen

Looking Ahead: *Sixth Grade Everyday Mathematics*

Next year your child will ...

◆ continue to collect, display, describe, and interpret data.

◆ maintain and extend skills for comparing, adding, subtracting, multiplying, and dividing fractions and mixed numbers.

◆ use scientific notation to write large and small numbers, and explore scientific notation on a calculator.

◆ continue the study of variables, expressions, equations, and other topics in algebra; use variables in spreadsheets; and solve equations and inequalities.

◆ extend skills in geometry, including constructions, transformations of figures, and finding volumes of 3-dimensional figures.

◆ maintain and apply skills for adding, subtracting, multiplying, and dividing whole numbers, decimals, and positive and negative numbers.